CONFORMAL
REPRESENTATION

CONFORMAL REPRESENTATION

C. CARATHÉODORY

DOVER PUBLICATIONS, INC.
Mineola, New York

Bibliographical Note

This Dover edition, first published in 1998, is an unabridged republication of the work first issued in 1932 and again in 1952 in a revised edition by Cambridge University Press, London.

Library of Congress Cataloging-in-Publication Data

Carathéodory, Constantin, 1873–1950.
    Conformal representation / C. Carathéodory.
        p. cm.
    Originally published: 2nd ed. Cambridge : University Press, 1952.
(Cambridge tracts in mathematics and mathematical physics; no. 28)
    Includes bibliographical references (p. –) and index.
    ISBN 0-486-40028-X (pbk.)
    1. Conformal mapping. 2. Geometry, Non-Euclidian. 3. Surfaces,
Representation of. I. Title.
QA360.C3   1998
516.3'6—dc21
                                                 97-47129
                                                    CIP

Manufactured in the United States of America
Dover Publications, Inc., 31 East 2nd Street, Mineola, N.Y. 11501

NOTE BY THE GENERAL EDITOR

PROFESSOR Carathéodory made the few corrections necessary in the text of the first edition of this tract, completed the manuscript of the new Chapter VIII, compiled the Bibliographical Notes afresh, and wrote the Preface to the second edition, during the later months of 1949. At his request, Mr G. E. H. Reuter, of the University of Manchester, agreed to revise the author's English. This work was completed shortly before Professor Carathéodory's death on 2 February 1950. Mr Reuter then kindly undertook the task of reading the proofs and attended to all points of detail which arose while the tract was going through the Press.

W.V.D.H.

CAMBRIDGE
January 1951

PREFACE TO SECOND EDITION

THIS reprint of my tract is almost without change, save for the addition of a chapter on the celebrated theorem of Poincaré and Koebe on uniformisation. If I have succeeded in making this chapter rather short, it is because I have been able to avail myself of the beautiful proof of van der Waerden (30), which has enabled me to expound the topological side of the problem in a few pages.

C. CARATHÉODORY

MUNICH
December 1949

PREFACE TO THE FIRST EDITION

THIS little book is the outcome of lectures which I have given at various times and at different places (Göttingen, Berlin, Athens, Munich, and at the University of Harvard). It contains the theory of conformal representation as it has developed during the last two decades. The first half of the book deals with some elementary subjects, knowledge of which is essential for the understanding of the general theory. The exposition of this theory in the last three chapters uses the simplest methods available to-day.

The original manuscript, written in German, has been translated by Mr B. M. Wilson of the University of Liverpool and by Miss Margaret Kennedy of Newnham College. I wish to express here my warmest thanks for the care they have taken that the most intricate arguments should be made clear to the reader. I am also indebted to Prof. Erhard Schmidt (Berlin) and to Prof. Tibor Radó (Columbus, Ohio) for various improvements in the mathematical demonstrations, and to Miss Kennedy for several suggestions that simplified the text. Finally, my thanks are due to the staff of the Cambridge University Press for the admirable way in which their part of the work has been carried out.

C. CARATHÉODORY

ATHENS
December 1931

CONTENTS

INTRODUCTION

HISTORICAL SUMMARY

1. By an isogonal (*winkeltreu*) representation of two areas on one another we mean a one-one, continuous, and continuously differentiable representation of the areas, which is such that two curves of the first area which intersect at an angle α are transformed into two curves intersecting at the same angle α. If the sense of rotation of a tangent is preserved, an isogonal transformation is called *conformal*.

Disregarding as trivial the Euclidean magnification (*Ähnlichkeitstransformation*) of the plane, we may say that the oldest known transformation of this kind is the *stereographic projection* of the sphere, which was used by *Ptolemy* (flourished in the second quarter of the second century; died after A.D. 161) for the representation of the celestial sphere; it transforms the sphere conformally into a plane. A quite different conformal representation of the sphere on a plane area is given by *Mercator's Projection*; in this the spherical earth, cut along a meridian circle, is conformally represented on a plane strip. The first map constructed by this transformation was published by *Mercator* (1512–1594) in 1568, and the method has been universally adopted for the construction of sea-maps.

2. A comparison of two maps of the same country, one constructed by stereographic projection of the spherical earth and the other by Mercator's Projection, will show that conformal transformation does not imply similarity of corresponding figures. Other non-trivial conformal representations of a plane area on a second plane area are obtained by comparing the various stereographic projections of the spherical earth which correspond to different positions of the centre of projection on the earth's surface. It was considerations such as these which led *Lagrange* (1736–1813) in 1779 to obtain all conformal representations of a portion of the earth's surface on a plane area wherein all circles of latitude and of longitude are represented by circular arcs[1].

3. In 1822 *Gauss* (1777–1855) stated and completely solved the general problem of finding *all* conformal transformations which transform a sufficiently small neighbourhood of a point on an arbitrary analytic surface into a plane area(2). This work of Gauss appeared to give the whole inquiry its final solution; actually it left unanswered the much more difficult question whether and in what way a given finite portion of the surface can be represented on a portion of the plane. This was first pointed out by *Riemann* (1826–1866), whose Dissertation (1851) marks a turning-point in the history of the problem which has been decisive for its whole later development; Riemann not only introduced all the ideas which have been at the basis of all subsequent investigation of the problem of conformal representation, but also showed that the problem itself is of fundamental importance for the theory of functions(3).

4. Riemann enunciated, among other results, the theorem that every simply-connected plane area which does not comprise the whole plane can be represented conformally on the interior of a circle. In the proof of this theorem, which forms the foundation of the whole theory, he assumes as obvious that a certain problem in the calculus of variations possesses a solution, and this assumption, as *Weierstrass* (1815–1897) first pointed out, invalidates his proof. Quite simple, analytic, and in every way regular problems in the calculus of variations are now known which do not always possess solutions(4). Nevertheless, about fifty years after Riemann, *Hilbert* was able to prove rigorously that the particular problem which arose in Riemann's work does possess a solution; this theorem is known as *Dirichlet's Principle*(5).

Meanwhile, however, the truth of Riemann's conclusions had been established in a rigorous manner by *C. Neumann* and, in particular, by *H. A. Schwarz*(6). The theory which Schwarz created for this purpose is particularly elegant, interesting and instructive; it is, however, somewhat intricate, and uses a number of theorems from the theory of the logarithmic potential, proofs of which must be included in any complete account of the method. During the present century the work of a number of mathematicians has created new methods which make possible a very simple treatment of our problem; it is the purpose of the following pages to give an account of these methods which, while as short as possible, shall yet be essentially complete.

CHAPTER I

MÖBIUS TRANSFORMATION

5. Conformal representation in general.

It is known from the theory of functions that an analytic function $w = f(z)$, which is regular and has a non-zero differential coefficient at the point $z = z_0$, gives a continuous one-one representation of a certain neighbourhood of the point z_0 of the z-plane on a neighbourhood of a point w_0 of the w-plane.

Expansion of the function $f(z)$ gives the series

$$\left.\begin{aligned} w - w_0 = A(z - z_0) + B(z - z_0)^2 + \dots, \\ A \neq 0; \end{aligned}\right\} \qquad \dots\dots(5\cdot1)$$

and if we write

$$z - z_0 = re^{it}, \quad A = ae^{i\lambda}, \quad w - w_0 = \rho e^{iu}, \qquad \dots\dots(5\cdot2)$$

where t, λ, and u are real, and r, a, and ρ are positive, then $(5\cdot1)$ may be written

$$\left.\begin{aligned} \rho e^{iu} = are^{i(\lambda + t)}\{1 + \phi(r, t)\}, \\ \lim_{r \to 0} \phi(r, t) = 0. \end{aligned}\right\} \qquad \dots\dots(5\cdot3)$$

This relation is equivalent to the following two relations:

$$\left.\begin{aligned} \rho = ar\{1 + \alpha(r, t)\}, \quad u = \lambda + t + \beta(r, t), \\ \lim_{r \to 0} \alpha(r, t) = 0, \qquad \lim_{r \to 0} \beta(r, t) = 0. \end{aligned}\right\} \qquad \dots\dots(5\cdot4)$$

When $r = 0$ the second of these relations becomes

$$u = \lambda + t, \qquad \dots\dots(5\cdot5)$$

and expresses the connection between the direction of a curve at the point z_0 and the direction of the corresponding curve at the point w_0. Equation $(5\cdot5)$ shows in particular that the representation furnished by the function $w = f(z)$ at the point z_0 is isogonal. Since the derivative $f'(z)$ has no zeros in a certain neighbourhood of z_0, it follows that the representation effected by $f(z)$ of a neighbourhood of z_0 on a portion of the w-plane is not only continuous but also conformal.

The first of the relations $(5\cdot4)$ can be expressed by saying that "infinitely small" circles of the z-plane are transformed into infinitely small circles of the w-plane. Non-trivial conformal transformations exist however for which this is also true of *finite* circles; these transformations will be investigated first.

6. Möbius Transformation.

Let A, B, C denote three real or complex constants, $\overline{A}$, $\overline{B}$, $\overline{C}$ their conjugates, and x, $\overline{x}$ a complex variable and its conjugate; then the equation

$$(A + \overline{A}) x\overline{x} + Bx + \overline{B}\overline{x} + C + \overline{C} = 0 \qquad \ldots\ldots(6\text{·}1)$$

represents a real circle or straight line provided that

$$B\overline{B} > (A + \overline{A})(C + \overline{C}). \qquad \ldots\ldots(6\text{·}2)$$

Conversely every real circle and every real straight line can, by suitable choice of the constants, be represented by an equation of the form (6·1) satisfying condition (6·2). If now in (6·1) we make any of the substitutions

$$y = x + \lambda, \qquad\qquad \ldots\ldots(6\text{·}3)$$

$$y = \mu x, \qquad\qquad \ldots\ldots(6\text{·}4)$$

or

$$y = \frac{1}{x}, \qquad\qquad \ldots\ldots(6\text{·}5)$$

the equation obtained can be brought again into the form (6·1), with new constants A, B, C which still satisfy condition (6·2). The substitution (6·5) transforms those circles and straight lines (6·1) for which $C + \overline{C} = 0$, i.e. those which pass through the point $x = 0$, into straight lines; we shall therefore regard straight lines as circles which pass through the point $x = \infty$.

7. If we perform successively any number of transformations (6·3), (6·4), (6·5), taking each time arbitrary values for the constants λ, μ, the resulting transformation is always of the form

$$y = \frac{\alpha x + \beta}{\gamma x + \delta}; \qquad\qquad \ldots\ldots(7\text{·}1)$$

here α, β, γ, δ are constants which necessarily satisfy the condition

$$\alpha\delta - \beta\gamma \neq 0, \qquad\qquad \ldots\ldots(7\text{·}2)$$

since otherwise the right-hand member of (7·1) would be either constant or meaningless, and (7·1) would not give a transformation of the x-plane. Conversely, any bilinear transformation (7·1) can easily be obtained by means of transformations (6·3), (6·4), (6·5), and hence (7·1) also transforms circles into circles.

The transformation (7·1) was first studied by *Möbius* (7) (1790–1868), and will therefore be called *Möbius' Transformation*.

8. The transformation inverse to (7·1), namely

$$x = \frac{-\delta y + \beta}{\gamma y - \alpha}, \qquad (-\delta)(-\alpha) - \beta\gamma \neq 0, \qquad \ldots\ldots(8\text{·}1)$$

is also a Möbius' transformation. Further, if we perform first the transformation (7·1) and then a second Möbius' transformation

$$z = \frac{\alpha_1 y + \beta_1}{\gamma_1 y + \delta_1}, \quad \alpha_1 \delta_1 - \beta_1 \gamma_1 \neq 0,$$

the result is a third Möbius' transformation

$$z = \frac{\mathrm{A}x + \mathrm{B}}{\Gamma x + \Delta},$$

with non-vanishing determinant

$$\mathrm{A}\Delta - \mathrm{B}\Gamma = (\alpha\delta - \beta\gamma)(\alpha_1\delta_1 - \beta_1\gamma_1) \neq 0.$$

Thus we have the theorem: *the aggregate of all Möbius' transformations forms a group.*

9. Equations (7·1) and (8·1) show that, if the x-plane is closed by the addition of the point $x = \infty$, *every Möbius' transformation is a one-one transformation of the closed x-plane into itself.* If $\gamma \neq 0$, the point $y = \alpha/\gamma$ corresponds to the point $x = \infty$, and $y = \infty$ to $x = -\delta/\gamma$; but if $\gamma = 0$ the points $x = \infty$ and $y = \infty$ correspond to each other.

From (7·1) we obtain

$$\frac{dy}{dx} = \frac{\alpha\delta - \beta\gamma}{(\gamma x + \delta)^2},$$

so that, by § 5, the representation is conformal except at the points $x = \infty$ and $x = -\delta/\gamma$. In order that these two points may cease to be exceptional we now extend the definition of conformal representation as follows: a function $y = f(x)$ will be said to transform the neighbourhood of a point x_0 conformally into a neighbourhood of $y = \infty$ if the function $\eta = 1/f(x)$ transforms the neighbourhood of x_0 conformally into a neighbourhood of $\eta = 0$; also $y = f(x)$ will be said to transform the neighbourhood of $x = \infty$ conformally into a neighbourhood of y_0 if

$$y = \phi(\xi) = f(1/\xi)$$

transforms the neighbourhood of $\xi = 0$ conformally into a neighbourhood of y_0. In this definition y_0 may have the value ∞.

In virtue of the above extensions we now have the theorem: *every Möbius' transformation gives a one-one conformal representation of the entire closed x-plane on the entire closed y-plane.*

10. Invariance of the cross-ratio.

Let x_1, x_2, x_3, x_4 denote any four points of the x-plane, and y_1, y_2, y_3, y_4 the points which correspond to them by the Möbius' transformation (7·1). If we suppose in the first place that all the numbers x_i, y_i

are finite, we have, for any two of the points,

$$y_k - y_i = \frac{\alpha x_k + \beta}{\gamma x_k + \delta} - \frac{\alpha x_i + \beta}{\gamma x_i + \delta} = \frac{\alpha\delta - \beta\gamma}{(\gamma x_k + \delta)(\gamma x_i + \delta)}(x_k - x_i),$$

and consequently, for all four,

$$\frac{(y_1 - y_4)(y_3 - y_2)}{(y_1 - y_2)(y_3 - y_4)} = \frac{(x_1 - x_4)(x_3 - x_2)}{(x_1 - x_2)(x_3 - x_4)}. \qquad \ldots\ldots(10\cdot1)$$

The expression

$$\frac{(x_1 - x_4)(x_3 - x_2)}{(x_1 - x_2)(x_3 - x_4)}$$

is called the cross-ratio of the four points x_1, x_2, x_3, x_4, so that, by (10·1), the cross-ratio is invariant under any Möbius' transformation.

A similar calculation shows that equation (10·1), when suitably modified, is still true if one of the numbers x_i or one of the numbers y_i is infinite; if, for example, $x_2 = \infty$ and $y_1 = \infty$,

$$\frac{y_3 - y_2}{y_3 - y_4} = \frac{x_1 - x_4}{x_3 - x_4}. \qquad \ldots\ldots(10\cdot2)$$

11. Let x_1, x_2, x_3 and y_1, y_2, y_3 be two sets each containing three unequal complex numbers. We will suppose in the first place that all six numbers are finite. The equation

$$\frac{(y_1 - y)(y_3 - y_2)}{(y_1 - y_2)(y_3 - y)} = \frac{(x_1 - x)(x_3 - x_2)}{(x_1 - x_2)(x_3 - x)} \qquad \ldots\ldots(11\cdot1)$$

when solved for y yields a Möbius' transformation which, as is easily verified, transforms each point x_i into the corresponding point y_i*, and § 10 now shows that it is the *only* Möbius' transformation which does so. This result remains valid when one of the numbers x_i or y_i is infinite, provided of course that equation (11·1) is suitably modified.

12. Since a circle is uniquely determined by three points on its circumference, § 11 may be applied to find Möbius' transformations which transform a given circle into a second given circle or straight line. Thus, for example, by taking $x_1 = 1$, $x_2 = i$, $x_3 = -1$ and $y_1 = 0$, $y_2 = 1$, $y_3 = \infty$, we obtain the transformation

$$y = i\frac{1 - x}{1 + x}, \qquad \ldots\ldots(12\cdot1)$$

i.e. one of the transformations which represent the circle $|x| = 1$ on the real axis, and the *interior* $|x| < 1$ of the unit-circle on the *upper* half of the y-plane. By a different choice of the six points x_i, y_i we can represent the exterior $|x| > 1$ of the unit-circle on this same half-plane.

_____

* The determinant of this transformation has the value

$$\alpha\delta - \beta\gamma = (y_1 - y_2)(y_1 - y_3)(y_2 - y_3)(x_1 - x_2)(x_1 - x_3)(x_2 - x_3).$$

In particular by taking the three points y_i on the same circle as the points x_i we can transform the interior of this circle into itself or into the exterior of the circle according as the points x_1, x_2, x_3 and y_1, y_2, y_3 determine the same or opposite senses of description of the perimeter. If, for example, in (11·1) we put $y_1 = 0$, $y_2 = 1$, $y_3 = \infty$, and then successively $x_1 = 1$, $x_2 = \infty$, $x_3 = 0$ and $x_1 = \infty$, $x_2 = 1$, $x_3 = 0$, we obtain the two transformations

$$y = (x-1)/x \quad \text{and} \quad y = 1/x; \qquad \ldots\ldots(12\text{·}2)$$

the first transforms the upper half-plane into itself, whereas the second transforms it into the lower half-plane.

13. Pencils of circles.

Since a Möbius' transformation is conformal it transforms orthogonal circles into orthogonal circles. We shall now show that, *given any two circles A and B, we can find a Möbius' transformation which transforms them either into two straight lines or into two concentric circles.*

If A and B have at least one common point P, then any Möbius' transformation whereby P corresponds to the point ∞ transforms A and B into straight lines; these lines intersect or are parallel according as A and B have a common point other than P, or not.

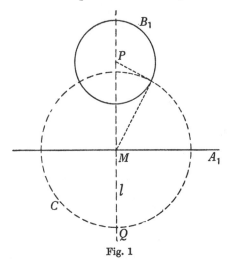

Fig. 1

If A and B have no common point, first transform the circle A by a Möbius' transformation into a straight line A_1, and let B_1 be the circle corresponding to B; A_1 and B_1 do not intersect. Draw the straight line l through the centre of B_1 perpendicular to A_1; let the foot of this perpendicular be M. With centre M draw the circle C cutting B_1 orthogonally. By a second Möbius' transformation we can transform the circle C and the straight line l into two (orthogonal) intersecting straight lines; A_1, B_1 are thereby transformed into two circles A_2, B_2, which cut both these straight lines orthogonally and are therefore *concentric*.

14. Given two intersecting straight lines there is a family of concentric circles orthogonal to both; given two parallel straight lines there is a family of parallel straight lines orthogonal to both; and given two concentric circles there is a family of intersecting straight lines orthogonal to both. Each of these families of circles or straight lines consists of all circles (or straight lines) of the plane which cut both the given lines or circles orthogonally. Since a Möbius' transformation is isogonal it follows that: *given any two circles A, B, there exists exactly one one-parametric family of circles which cut A and B orthogonally; this family is called the pencil of circles conjugate to the pair A, B.*

If the circles A and B intersect in two points P, Q of the plane, no two circles of the conjugate pencil can intersect, and the pencil is then said to be *elliptic*. No circle of the pencil passes through either of the points P, Q, which are called the *limiting points* of the pencil.

Secondly, if A and B touch at a point P, the conjugate pencil consists of circles all of which touch at P, and is called *parabolic*; P is the *common point (Knotenpunkt)* of the pencil.

Lastly, if A and B have no point in common, the conjugate pencil consists of all circles which pass through two fixed points, the *common points* of the pencil, and is called *hyperbolic*.

15. Considering the three types of pencils of circles as defined in § 14, we see that if C, D are any two circles of the pencil conjugate to A, B, then A, B belong to the pencil conjugate to C, D. This pencil containing A, B is independent of the choice of the two circles C, D, and we therefore have the following theorem : *there is one and only one pencil of circles which contains two arbitrarily given circles; i.e. a pencil of circles is uniquely determined by any two of its members.*

We see further from the three standard forms of pencils that : *through every point of the plane which is neither a limiting point nor a common point of a given pencil of circles there passes exactly one circle of the pencil.*

16. Bundles of circles.

Let A, B, C be three circles which do not all pass through a common point P. If A, B have no common point we can transform them (§ 13) by a Möbius' transformation into concentric circles A_1, B_1, and that common diameter of A_1 and B_1 which cuts C_1 (the circle into which C is transformed) orthogonally is a circle of the plane cutting all three circles A_1, B_1, C_1 orthogonally. Hence a circle exists which cuts all three circles A, B, C orthogonally.

Secondly, if A and B touch, there is a Möbius' transformation which transforms them into two parallel straight lines, and C into a circle C_1. Since C_1 has one diameter perpendicular to the two parallel straight lines, a circle exists in this case also cutting all three circles A, B, C orthogonally.

Finally, if A and B have two points in common, there is a Möbius' transformation which transforms them into two straight lines intersecting at a point O, and C into a circle C_1 which does not pass through O. Two cases must now be distinguished: if O lies *outside* the circle C_1 there is

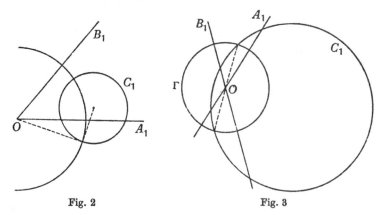

Fig. 2   Fig. 3

again a circle cutting A_1, B_1, and C_1 orthogonally; whereas if O lies *inside* C_1 there is a circle Γ such that each of the circles A_1, B_1, C_1 intersects Γ at the extremities of a diameter of Γ.

We have thus proved the following theorem: *any three co-planar circles must satisfy at least one of the following conditions: the three circles have a common orthogonal circle K, or they pass through a common point, or they can be transformed by a Möbius' transformation into three circles which cut a fixed circle Γ at the extremities of a diameter of Γ.* It follows readily from the proof given that if the three circles A, B, C do not belong to the same pencil the circle K is unique; further, it will be proved below that three given circles *cannot satisfy more than one* of the three conditions enumerated.

17. We now define three types of families of circles which we call *bundles of circles.*

An *elliptic bundle of circles* consists of all circles of the plane which cut a fixed circle Γ at the extremities of a diameter of Γ. The circle Γ itself belongs to the bundle and is called the *equator* of the bundle.

A *parabolic bundle of circles* consists of all circles of the plane which pass through a fixed point, the *common point* of the bundle.

A *hyperbolic bundle of circles* consists of all circles of the plane which cut a fixed circle or straight line orthogonally.

These three figures are *essentially* distinct: every pair of circles of an elliptic bundle intersect at two points; every pair of circles of a parabolic bundle either intersect at two points or touch one another; but a hyperbolic bundle contains pairs of circles which have no common point.

18. Bundles of circles nevertheless possess very remarkable common properties. For example: *if A, B are two circles of a bundle, all the circles of the pencil which contains A, B belong to this bundle.* For a parabolic bundle the truth of this theorem is obvious; for a hyperbolic bundle it follows from the fact that the orthogonal circle of the bundle cuts the circles A, B—and therefore cuts every circle of the pencil containing A, B—orthogonally; and for an elliptic bundle it follows from an elementary theorem of Euclid.

The proof of the following theorem is equally simple: *if a plane contains a bundle of circles and an arbitrary point P, which, if the bundle is parabolic, does not coincide with the common point of the bundle, then P lies on an infinite number of circles of the bundle, and these circles through P form a pencil.*

19. Let A, B, C be three circles of a bundle which do not belong to the same pencil, and let D be any fourth circle of the bundle; then, starting with A, B, C we can, by successive construction of pencils, arrive at a pencil of circles which contains D, and all of whose members are circles of the bundle. For there is on D at least one point P which is neither a common point nor a limiting point of either of the two pencils determined by A, B and by A, C and which does not lie on A; we can therefore draw through P two circles E, F, so that E belongs to the pencil A, B, and F to the pencil A, C. The circles E, F are distinct, since A, B, C do not belong to the same pencil, and the second theorem of § 18 now shows that D belongs to the pencil determined by E, F.

It follows that a bundle of circles is uniquely determined by any three of its members which do not belong to the same pencil, and in particular that three circles of an elliptic bundle which do not belong to the same pencil cannot have a common orthogonal circle; for if they had they would define a bundle that was both elliptic and hyperbolic.

20. The circles obtained by applying a Möbius' transformation to all the circles of a bundle also form a bundle, and the two bundles are of the same kind.

For parabolic and hyperbolic bundles this theorem is an immediate consequence of the definitions of these figures. We therefore denote by M the aggregate of circles obtained from the circles of a given elliptic bundle by means of a Möbius' transformation; all those circles of M which pass through the point ∞ form a pencil of straight lines, intersecting at a point O of the plane. Let A, B be any two straight lines through O, and let C be any third circle of M. Since, by § 19, the circles A, B, C cannot have a common orthogonal circle, the point O must be interior to C, and consequently, by § 16, the circles A, B, C belong to an elliptic bundle, the circles of which can all be obtained from A, B, C by successive construction of pencils. And this bundle must be identical with M, since M is obtained by precisely the same construction.

21. This last result, together with § 16, shows that any three co-planar circles which do not belong to the same pencil determine exactly one bundle.

22. A bundle of circles cannot contain an elliptic pencil together with its conjugate hyperbolic pencil. For since, by § 17, neither an elliptic nor a parabolic bundle can contain an elliptic pencil, the given bundle would necessarily be *hyperbolic*, so that there would be a circle, the orthogonal circle of the bundle, cutting all members of the two given conjugate pencils orthogonally; but this is impossible.

23. Inversion with respect to a circle.

Given a straight line a and a point P, let P^* be the image-point of P in a; we shall call P^* the *inverse* point of P with respect to a. More generally, given a fixed circle A, we can, by a Möbius' transformation, transform A into a straight line; consequently for every point P there is a point P^* inverse to P with respect to A, and P^* is characterised by the fact that every circle through P and orthogonal to A also passes through P^*.

The operation of inversion is involutory; further, the figure formed by a circle A and two inverse points is transformed by any Möbius' transformation into a circle A and two inverse points. Thus, since inversion with respect to a straight line gives an isogonal but not a conformal representation of the plane on itself, inversion with respect to a circle A does so also.

24. If t is an arbitrary point of the complex t plane, its inverse with respect to the real axis is given by the conjugate complex number $\bar{t}$; more generally, the points

$$x = e^{i\theta}t, \quad y = e^{i\theta}\bar{t} \qquad \ldots\ldots(24\cdot1)$$

are inverse points with respect to the straight line through the origin obtained by rotating the real axis through an angle θ. The first of equations (24·1) gives $\bar{x} = e^{-i\theta}\bar{t}$, so that

$$y = e^{2i\theta}\bar{x}. \qquad \ldots\ldots(24\cdot2)$$

Similarly, since the equation

$$x = \frac{a(1+it)}{1-it} \qquad \ldots\ldots(24\cdot3)$$

transforms the real axis of the t-plane into the circle $|x| = a$, it follows that the points

$$x = \frac{a(1+it)}{1-it}, \quad y = \frac{a(1+i\bar{t})}{1-i\bar{t}} \qquad \ldots\ldots(24\cdot4)$$

are inverse points with respect to $|x| = a$; from this we at once deduce that

$$y = a^2/\bar{x}. \qquad \ldots\ldots(24\cdot5)$$

25. Two successive inversions are equivalent either to a Möbius' transformation or to the identical transformation. If, for example, the inversions are performed with respect to the straight lines through the origin O which make angles θ, $\theta + \phi$ with the real axis, we have, by (24·2),

$$y = e^{2i\theta}\bar{x}, \quad z = e^{2i(\theta+\phi)}\bar{y},$$

and consequently

$$z = e^{2i\phi}x. \qquad \ldots\ldots(25\cdot1)$$

The resulting transformation is therefore a rotation of the plane about the origin through an angle 2ϕ; thus the angle of rotation depends only on the angle between the two given straight lines, *not on the position of these lines.*

Similarly, for two inversions with respect to the concentric circles $|x| = a$ and $|x| = b$, we have, by (24·5),

$$y = a^2/\bar{x}, \quad z = b^2/\bar{y},$$

and so

$$z = \frac{b^2}{a^2}x. \qquad \ldots\ldots(25\cdot2)$$

The resulting transformation is therefore a *magnification (Ähnlichkeits-*

transformation) which depends only on the ratio $b : a$ of the radii, not on the lengths of the radii themselves.

Similarly it is seen that two successive inversions with respect to parallel straight lines are equivalent to a translation, which depends only on the direction of the parallel lines and the distance between them, but not on their position in the plane.

Since, by §13, any two circles can, by means of a suitable Möbius' transformation, be transformed into one of the above three figures, we have proved the following theorem: *two successive inversions of the plane with respect to an arbitrary pair of circles A, B are equivalent to a Möbius' transformation; the same transformation is obtained by successive inversions with respect to two other suitable circles A_1, B_1 of the pencil defined by A, B; and one of the two circles A_1, B_1 may be taken arbitrarily in this pencil. Further: the resulting Möbius' transformation leaves all circles of the pencil conjugate to A, B invariant.*

26. Geometry of Möbius Transformations.

Points of the complex plane exist which are invariant for the transformation

$$y = \frac{\alpha x + \beta}{\gamma x + \delta}. \qquad \ldots\ldots(26\cdot1)$$

For these points $y = x$; i.e. they are the roots of the equation

$$\gamma x^2 + (\delta - \alpha) x - \beta = 0. \qquad \ldots\ldots(26\cdot2)$$

If all the coefficients in this equation vanish, the given transformation is the identical one $y = x$, and every point of the plane is a fixed point.

If $\gamma \neq 0$, let x_1, x_2 be the roots of (26·2), so that

$$\left. \begin{aligned} x_i &= \frac{\alpha - \delta \pm \sqrt{D}}{2\gamma}, \\ D &= (\alpha + \delta)^2 - 4(\alpha\delta - \beta\gamma); \end{aligned} \right\} \qquad \ldots\ldots(26\cdot3)$$

thus the number of fixed points is one or two according as $D = 0$ or $D \neq 0$.

If $\gamma = 0$, the point $x = \infty$ is to be regarded as a fixed point, so that in this case also the number of fixed points is one or two according as $D = 0$ or $D \neq 0$.

27. Suppose first that $D = 0$. If also $\gamma = 0$, then $\alpha = \delta$, and (26·1) is of the form

$$y = x + \frac{\beta}{\alpha}. \qquad \ldots\ldots(27\cdot1)$$

This is a translation and can be obtained by two inversions with respect to parallel straight lines. If however $\gamma \neq 0$, (26·3) gives

$$x_1 = x_2 = \frac{\alpha - \delta}{2\gamma}. \qquad \ldots\ldots(27\cdot2)$$

But on solving the equation

$$\frac{1}{y - x_1} = \frac{1}{x - x_1} + \frac{2\gamma}{\alpha + \delta} \qquad \ldots\ldots(27\cdot3)$$

for y, we obtain a Möbius' transformation identical with (26·1)*. If we now introduce new coordinates

$$\omega = 1/(y - x_1), \quad t = 1/(x - x_1),$$

the transformation again becomes a *translation*, namely

$$\omega = t + \frac{2\gamma}{\alpha + \delta}.$$

We therefore have the following theorem: *any Möbius' transformation* (26·1) *for which the discriminant D is zero can be obtained by two successive inversions with respect to two circles which touch each other.*

28. Suppose secondly that $D \neq 0$. If $\gamma = 0$, then $\alpha \neq \delta$; and by putting

$$\omega = y - \frac{\beta}{\delta - \alpha}, \quad t = x - \frac{\beta}{\delta - \alpha},$$

we obtain, on elimination of x and y,

$$\omega = \frac{\alpha}{\delta} t. \qquad \ldots\ldots(28\cdot1)$$

If on the other hand $\gamma \neq 0$, write

$$\omega = \frac{y - x_2}{y - x_1}, \quad t = \frac{x - x_2}{x - x_1}; \qquad \ldots\ldots(28\cdot2)$$

with the new coordinates the points $t = 0$, $t = \infty$ must be fixed points of the transformation, which must therefore be of the form

$$\omega = \rho t. \qquad \ldots\ldots(28\cdot3)$$

Corresponding to the point $x = \infty$ we have $y = \alpha/\gamma$, $t = 1$; thus (28·3) shows that

$$\rho = \omega = \frac{\alpha - \gamma x_2}{\alpha - \gamma x_1}.$$

This equation may also be written

$$\rho = \frac{\alpha + \delta + \sqrt{D}}{\alpha + \delta - \sqrt{D}}. \qquad \ldots\ldots(28\cdot4)$$

---

* This is seen, for example, by noting that both transformations transform the points $x' = x_1$, $x'' = \infty$, $x''' = -\delta/\gamma$ into the points $y' = x_1$, $y'' = a/\gamma$, $y''' = \infty$.

Since, if $\gamma = 0$, equations (28·3) and (28·4) reduce to (28·1), the case when $\gamma = 0$ need no longer be treated separately.

29. If ρ is real and positive, the transformation $\omega = \rho t$ is a magnification and can be obtained by two inversions with respect to concentric circles.

Secondly, if $|\rho| = 1$, i.e. if $\rho = e^{i\theta}$, the transformation is a rotation of the plane, and can be obtained by two inversions with respect to two intersecting straight lines.

If neither of these conditions is satisfied, then $\rho = ae^{i\theta}$, where $\theta \not\equiv 0$ (mod 2π), $a > 0$, and $a \neq 1$. The transformation can be obtained by a rotation followed by a magnification, i.e. by four successive inversions with respect to circles. Since, as is easily seen, no circle is transformed into itself, it is not possible to obtain the transformation by two inversions only. The transformation is in this case said to be *loxodromic*.

30. The various cases can be clearly distinguished by introducing the parameter

$$\lambda = \frac{(\alpha + \delta)^2}{4(\alpha\delta - \beta\gamma)}. \qquad \ldots\ldots(30\cdot1)$$

Equation (28·4) now takes the form

$$\rho = \frac{\sqrt{\lambda} + \sqrt{(\lambda - 1)}}{\sqrt{\lambda} - \sqrt{(\lambda - 1)}},$$

and on solving this for λ we obtain

$$\lambda = \frac{(\rho + 1)^2}{4\rho} = 1 + \frac{(\rho - 1)^2}{4\rho}.$$

If now ρ is real, positive and different from unity, λ is real and greater than unity; secondly, if $\rho = e^{i\theta}$ ($\theta \not\equiv 0$ mod 2π), $\lambda = \cos^2 \frac{1}{2}\theta$, i.e. λ is real, positive and less than unity; finally $\lambda = 1$ if $D = 0$. Thus: *the transformation is always loxodromic if λ is not real or if λ is real and negative. If λ is real and positive, the transformation can be obtained by inversions with respect to two circles of an elliptic, parabolic, or hyperbolic pencil according as $\lambda > 1$, $\lambda = 1$, or $\lambda < 1$.*

CHAPTER II

NON-EUCLIDEAN GEOMETRY

31. Inversion with respect to the circles of a bundle.

We shall now consider the aggregate of Möbius' transformations which are obtained by two successive inversions with respect to circles of a given bundle. It will first be proved that these transformations form a group.

In the first place, the transformation *inverse* to any one of the transformations considered is obtained by inverting the order of the two inversions; for four inversions with respect to the circles A, B, B, A, taken in the order indicated, clearly produce the identical transformation.

32. To prove that the transformations considered form a group, it remains to prove that four successive inversions with respect to the circles A, B, C, D of the bundle can be replaced by two inversions with respect to circles of the same bundle.

Suppose in the first place that the two circles A, B intersect in two points P, Q, thus defining a hyperbolic pencil. It follows from § 22 that P and Q cannot both be limiting points of the pencil determined by C and D, and consequently there is at least one circle C_1 of this pencil passing through one of the points P, Q; suppose C_1 passes through P. By §18 the circles A, B, C_1 belong to the same pencil, and consequently (§25) the two inversions with respect to A, B can be replaced by inversions with respect to A_1, C_1, where A_1 is a circle of this pencil. Similarly, the inversions with respect to C, D can be replaced by inversions with respect to C_1, D_1, where D_1 is a circle of the pencil determined by C, D. The four inversions with respect to A, B, C, D are therefore equivalent to four inversions with respect to A_1, C_1, C_1, D_1, i.e. to two inversions with respect to A_1, D_1, since the two successive inversions with respect to C_1 destroy one another.

In the general case the pencil determined by B, C contains at least one circle B_1 which either coincides with A or intersects A at two points. The inversions with respect to B, C may now be replaced by inversions with respect to B_1, C_1, and consequently the inversions with respect to A, B, C, D by inversions with respect to A, B_1, C_1, D; then either the inversions with respect to A, B_1 destroy one another or the problem has been reduced to the case already dealt with.

The groups of transformations, the existence of which has just been established, are fundamentally distinct according as the bundle considered is elliptic, parabolic, or hyperbolic.

33. The circles of an elliptic bundle can be obtained by stereographic projection of the great circles of a suitable sphere. Any inversion of the plane with respect to a circle of the bundle corresponds to an ordinary inversion of the sphere with respect to the plane of the corresponding great circle, and the group of Möbius' transformations obtained is isomorphic with the group of rotations of the sphere.

34. The circles of a parabolic bundle can be transformed by a suitable Möbius' transformation into the aggregate of straight lines in the plane. Since every ordinary motion of the plane can be obtained by successive inversions with respect to two straight lines, the group of transformations is now isomorphic with the group of motions of a rigid plane.

35. Representation of a circular area on itself.

The most important case for us is that in which the bundle is *hyperbolic*, so that the group considered is obtained by inversions with respect to two circles which cut a given circle (or straight line) orthogonally.

It is seen immediately that for each single inversion, and therefore for every operation of the group, the circumference of the orthogonal circle is transformed into itself, and the interior of this circle into itself[*]. We will now prove that, *conversely, every Möbius' transformation of which this is true is a transformation of the group.* It is convenient to prove this theorem step by step; we shall suppose that the circular area which is transformed into itself is the unit-circle $|z| \leqslant 1$.

We first prove that any interior point Q of this area, with coordinate $z = a\,(|a| < 1)$, can be transformed into the origin P, $z = 0$, by a transformation of the group. The point Q_1 inverse to Q with respect to the unit-circle O has coordinate $1/\bar{a}$, and the circle C with centre $1/\bar{a}$ and radius $\sqrt{\left(\dfrac{1}{a\bar{a}} - 1\right)}$ is orthogonal to O. Now invert first with respect to the circle C and then with respect to the straight line PQ, both

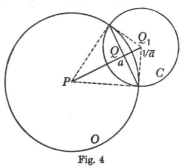

Fig. 4

---

[*] If the orthogonal circle is a straight line, each operation of the group transforms each of the half-planes defined by this straight line into itself.

of which are circles of the hyperbolic bundle considered. The first inversion transforms Q into P, and the second leaves P fixed, so that the final result is to transform Q into P.

The following observation will be of use later: the above Möbius' transformation transforms the points a, $a/\sqrt{(a\bar{a})}$, $1/\bar{a}$ into the points 0, $-a/\sqrt{(a\bar{a})}$, ∞ respectively, and is therefore (§ 11) given by the equation

$$y = \frac{a - x}{1 - \bar{a}x}. \qquad \ldots\ldots(35\cdot1)$$

Since rotations about the origin belong to the group considered, the result that has just been proved may be stated in the following sharper form: we can, by an operation of the group, transform the area $|z| \leqslant 1$ into itself in such a way that a directed line-element through an interior point of the area is transformed into a given directed line-element through the origin. In particular the transformation

$$y = \frac{x - a}{1 - \bar{a}x}, \qquad \ldots\ldots(35\cdot2)$$

being obtained from (35·1) by rotation through 180°, belongs to the group.

The complete theorem will have been proved if we show that a Möbius' transformation which represents the circle on itself in such a way that a given line-element through the origin (and consequently every line-element through the origin) is transformed into itself must of necessity be the identical transformation. But, for such a transformation, the points $x = 0$, $x = \infty$ are fixed points, and the point $x = 1$ is transformed into a point $x = e^{i\theta}$. The transformation must therefore be of the form $y = e^{i\theta}x$, and it follows from the invariance of line-elements through the origin that $\theta = 0$.

36. Non-Euclidean Geometry (8) (9).

The group of Möbius' transformations which represent a circular area (or a half-plane) on itself has many properties analogous to those of the group of motions of a rigid plane. In this comparison certain circular arcs in the circular area take the place of straight lines in the plane; these circular arcs are in fact the portions of the circles of the hyperbolic bundle which are inside the circular area or half-plane.

Thus, corresponding to the fact that a straight line in the Euclidean plane is determined uniquely by two points on it, we have the immediate theorem that through any two points in the upper half-plane (or inside the circle $|x| = 1$) one and only one circle can be drawn to cut the real axis (or the unit-circle) orthogonally. Again, it follows from what has already been said that one and only one circle of the hyperbolic bundle can be drawn through any given line-element in the upper half-plane (or in the circle $|x| < 1$); a precisely analogous statement holds for straight lines in the Euclidean plane.

In virtue of this analogy, the circular arcs in question will be called *non-Euclidean straight lines*, the half-plane (or circular area) *the non-Euclidean plane*, and the Möbius' transformations which transform the non-Euclidean plane into itself will be called *non-Euclidean motions*.

37. One fundamental contrast with ordinary geometry is however seen at once. For in Euclidean geometry, by Euclid's 11th postulate, through any point which does not lie on a given straight line one and only one straight line can be drawn which does not intersect the given straight line; whereas, in the non-Euclidean plane, through any point P which does not lie on a non-Euclidean straight line α an infinite number of non-Euclidean straight lines can be drawn none of which intersect α. Further, there are two non-Euclidean straight lines β and γ through P which divide all the remaining non-Euclidean straight lines through P into two classes: namely those which intersect α and those which do not. Two non-Euclidean straight lines such as α and β, which are in fact two circles which touch at a point of the orthogonal circle, are called *parallel* (*Lobatschewsky*, 1793–1856).

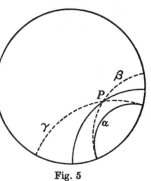

Fig. 5

38. Angle and distance.

Non-Euclidean motions, being Möbius' transformations which interchange the non-Euclidean straight lines, leave unaltered the angle between two intersecting straight lines, and consequently *the ordinary angle can also be taken as the non-Euclidean angle*.

The ordinary distance between two points, on the other hand, is not invariant for non-Euclidean motions, and an invariant function of two points must be determined to replace it if the idea of distance is to be employed in non-Euclidean geometry. Let

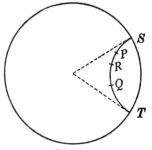

Fig. 6

P, Q be two points of a non-Euclidean plane, and let S, T denote the extremities of the non-Euclidean straight line joining P and Q. Then the cross-ratio λ of the four points (S, P, Q, T) is uniquely determined by the two points P, Q, and this cross-ratio is invariant for all non-Euclidean motions; hence any function $\psi(\lambda)$ of λ is also invariant. We shall now choose the function $\psi(\lambda)$ so that, if it is denoted by $D(P, Q)$, and if R is any point whatever of the segment PQ of the non-Euclidean straight line through P and Q, then

$$D(P, R) + D(R, Q) = D(P, Q). \qquad \ldots\ldots(38\cdot1)$$

39. Let h and Δh be two positive numbers whose sum is less than unity; denote by O the centre of the orthogonal circle $|x| = 1$, and by P, Q the points h, $h + \Delta h$. Then (38·1) gives

$$D(O, Q) = D(O, P) + D(P, Q). \qquad \ldots\ldots(39\cdot1)$$

Now the cross-ratio of the four points $-1, 0, h, 1$ is a function of h, so that we may write

$$D(O, P) = \phi(h), \quad D(O, Q) = \phi(h + \Delta h). \quad \ldots\ldots(39\cdot2)$$

To calculate $D(P, Q)$ we apply the non-Euclidean motion (§ 35)

$$y = \frac{x - h}{1 - hx};$$

this transforms P into O and Q into the point

$$\Delta h / \{1 - h(h + \Delta h)\},$$

and consequently, since distance is to be invariant,

$$D(P, Q) = \phi\left(\frac{\Delta h}{1 - h(h + \Delta h)}\right).$$

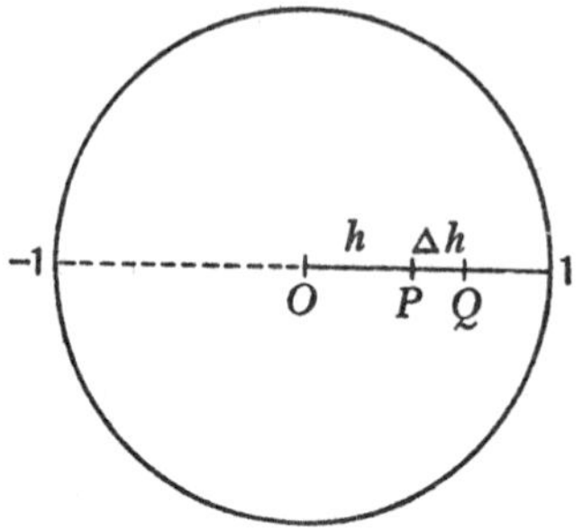

Fig. 7

Thus, by (39·1), the functional equation

$$\phi(h + \Delta h) = \phi(h) + \phi\left(\frac{\Delta h}{1 - h(h + \Delta h)}\right) \qquad \ldots\ldots(39\cdot3)$$

must be satisfied identically.

If we now assume that $\phi(u)$ is continuous and differentiable at $u = 0$ and that $\phi(0) = 0$, $\phi'(0) = 1$, (39·3) cannot hold unless

$$\lim_{\Delta h \to 0} \frac{\phi(h + \Delta h) - \phi(h)}{\Delta h} = \frac{1}{1 - h^2}.$$

It follows that $\phi(h)$ is differentiable for all values of h, that

$$\phi'(h) = \frac{1}{1 - h^2},$$

and hence, by integration, that

$$\phi(h) = \tfrac{1}{2} \log \frac{1 + h}{1 - h}. \qquad \ldots\ldots(39\cdot4)$$

Direct substitution in (39·3) then verifies that that equation is satisfied identically.

40. If z_1, z_2 are two arbitrary points of the non-Euclidean plane, the non-Euclidean motion (§ 35)

$$y = \frac{z - z_1}{1 - \bar{z}_1 z}$$

transforms z_1 into the point O, and z_2 into the point $(z_2 - z_1)/(1 - \bar{z}_1 z_2)$. By a rotation about O this last point can now be transformed into $|z_2 - z_1|/|1 - \bar{z}_1 z_2|$, so that, by (39·4),

$$D(z_1, z_2) = \tfrac{1}{2} \log \frac{|1 - \bar{z}_1 z_2| + |z_2 - z_1|}{|1 - \bar{z}_1 z_2| - |z_2 - z_1|};$$
$$|z_1| < 1, \quad |z_2| < 1. \qquad \qquad \dots\dots(40·1)$$

41. In the case when the orthogonal circle coincides with the real axis the non-Euclidean distance between two points z_1, z_2 in the upper half-plane can be deduced as follows: the Möbius' transformation

$$w = \frac{z - z_1}{\bar{z}_1 - z} \qquad \qquad \dots\dots(41·1)$$

transforms the real axis of the z-plane into the circle $|w| = 1$, the point z_1 into the point $w_1 = 0$, and the point z_2 into the point

$$w_2 = (z_2 - z_1)/(\bar{z}_1 - z_2).$$

In this way we obtain

$$D(z_1, z_2) = \tfrac{1}{2} \log \frac{|\bar{z}_1 - z_2| + |z_1 - z_2|}{|\bar{z}_1 - z_2| - |z_1 - z_2|};$$
$$\mathbf{I}(z_1) > 0, \quad \mathbf{I}(z_2) > 0. \qquad \qquad \dots\dots(41·2)$$

42. The triangle theorem.

It is easy to show by direct calculation from (40·1) that, for any three points of the non-Euclidean plane,

$$D(z_1, z_3) \leqslant D(z_1, z_2) + D(z_2, z_3), \qquad \dots\dots(42·1)$$

where the sign of equality holds if and only if z_2 lies on the non-Euclidean segment which joins z_1 and z_3. But the following proof of (42·1) is much more instructive.

All the axioms which Euclid postulates at the beginning of his work, excepting only the parallel postulate, hold also in the non-Euclidean plane. Hence, since the parallel postulate is not used in the earlier theorems, we can apply the first sixteen propositions of Euclid's first book without change to our figures; in particular, theorems concerning the congruence of triangles, the theorem that the greatest side of a triangle is opposite to the greatest angle, and, lastly, the triangle theorem expressed by (42·1).

43. Non-Euclidean length of a curve.

The conception of the length of a curve can at once be extended in the sense of this geometry to curves in the non-Euclidean plane. We suppose that the curve is given in terms of a parameter. On the curve take a finite set of points arranged in order so that the parameter always increases (or always decreases) from one point to the next. Let these points in order be taken as the corners of a polygon whose sides are non-Euclidean straight lines, and define the length of the polygon as the sum of the lengths of its sides; the upper limit (when finite) of the lengths of all such inscribed polygons now defines the length of the given curve.

Replacing z_1, z_2 in (40·1) by $z, z + \Delta z$, we obtain, as Δz tends to zero,

$$D(z, z + \Delta z) \sim \frac{|\Delta z|}{1 - |z^2|},$$

and it follows that the non-Euclidean length of an arc of the curve which is given by the complex function $z(t)$ is expressed by the integral

$$\int_{t_1}^{t_2} \frac{|z'(t)|}{1 - |z^2(t)|} \, dt.$$

The analogous formula for the case in which the orthogonal circle is the real axis is obtained in a similar way from (41·2); this formula is

$$\frac{1}{i} \int_{t_1}^{t_2} \frac{|z'|}{(\bar{z} - z)} \, dt.$$

44. Geodesic curvature.

It has been shown by *P. Finsler* (10) that, with the most general metric, the geodesic curvature of a curve C at a point P on it may be obtained as follows: consider any family of curves which contains C, and construct that geodesic line of the metric which touches C at P; let R be any point of this tangent geodesic, and let s denote the extremal distance PR and θ the angle of intersection* at R of the geodesic PR with that curve of the family which passes through R; then the geodesic curvature at P is

$$k = \lim_{s \to 0} \frac{\theta}{s}. \quad \dots\dots(44\cdot1)$$

It can be shown that the number k so obtained is independent of the particular choice of the family of curves.

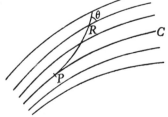

Fig. 8

* It is assumed that the measure of angles is suitably defined in the metric.

Applying the above definition of curvature to our non-Euclidean geometry, we see that curvature is invariant for non-Euclidean motions. Further, if the orthogonal circle is $|z| = 1$ and if, according to §39, $\phi'(0) = 1$, the curvature of a curve at the centre of this circle is the same as its ordinary Euclidean curvature, since the non-Euclidean straight lines through this point are ordinary straight lines and the value of angles is the same in both geometries. The non-Euclidean curvature of a curve $z(t)$ at a point z_0 may therefore be calculated by writing

$$\zeta(t) = \frac{z - z_0}{1 - z(t)\bar{z}_0}, \qquad \ldots\ldots(44\cdot2)$$

and then determining the ordinary curvature of $\zeta(t)$ at $\zeta = 0$. But this is known to be

$$k = \frac{\bar{\zeta}'\zeta'' - \zeta'\bar{\zeta}''}{2i(\zeta'\bar{\zeta}')^{\frac{3}{2}}}, \qquad \ldots\ldots(44\cdot3)$$

and consequently, from (44·2), we have, on replacing z_0 by z, the final formula

$$k = \frac{2z'\bar{z}'(\bar{z}z' - z\bar{z}') + (1 - z\bar{z})(\bar{z}'z'' - z'\bar{z}'')}{2i(z'\bar{z}')^{\frac{3}{2}}}. \qquad \ldots\ldots(44\cdot4)$$

45. Non-Euclidean motions.

It was shown in § 32 that every non-Euclidean motion can be obtained by two successive inversions with respect to non-Euclidean straight lines; there are three essentially distinct types of non-Euclidean motions, corresponding to different relative dispositions of these two straight lines.

Suppose first that the inversions are with respect to two non-Euclidean straight lines which intersect at a point P, where they make an angle $\frac{1}{2}\theta$ with one another; the motion is in this case a non-Euclidean rotation about P, the angle of rotation being θ. By keeping P fixed and letting θ vary we obtain a one-parameter group of non-Euclidean rotations.

Let P_1 be the point inverse to P with respect to the orthogonal circle; then every circle of the elliptic pencil which has the points P and P_1 as limiting points is transformed into itself by every operation of this group, so that the circles of this pencil, in so far as they lie inside the non-Euclidean plane, are non-Euclidean circles with common non-Euclidean centre P. If C is any one of these circles, all points on C are at the same non-Euclidean distance from P; this distance is the non-Euclidean radius of C.

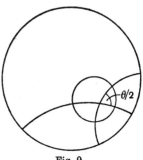

Fig. 9

46. Suppose, secondly, that the two inversions to which the motion is reducible are with respect to two non-Euclidean straight lines which neither intersect nor are parallel in the sense of Lobatschewsky. The conjugate pencil, consisting of all circles which are transformed into themselves by the motion, is hyperbolic; its common points M_1, M_2 lie on the orthogonal circle. Among the circles of this pencil there is exactly one non-Euclidean straight line. It intersects the non-Euclidean straight lines with respect to which the inversions are performed at the points P_1, P_2. If the non-Euclidean distance $P_1 P_2$ is denoted by $\frac{1}{2}h$, the motion can be regarded as a translation of the plane through distance h along the non-Euclidean straight line

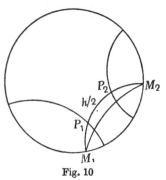

Fig. 10

$M_1 P_1 P_2 M_2$. By keeping the non-Euclidean straight line $M_1 M_2$ fixed and letting h vary we obtain a one-parameter group of motions, such that every motion of the group transforms into itself every circular arc which lies inside the non-Euclidean plane and joins the points M_1, M_2. These circular arcs are called *hypercycles*; a hypercycle can also be defined as the locus of a point whose non-Euclidean distance from the non-Euclidean straight line $M_1 M_2$ is constant.

Finally, if the two successive inversions are with respect to *parallel* non-Euclidean straight lines (§ 37), whose common end-point is L, we obtain a non-Euclidean motion which is called a *limit-rotation* (*Grenzdrehung*).

Given two pairs of parallel non-Euclidean straight lines LM, LN and $L_1 M_1$, $L_1 N_1$, there is always a non-Euclidean motion which transforms each pair into the other pair, so that the points L_1, M_1, N_1 or L_1, N_1, M_1. It follows that from the point of view of non-Euclidean geometry all limit-rotations are equivalent, or rather that they can differ only in the sense of the rotation. in the sense of the rotation.

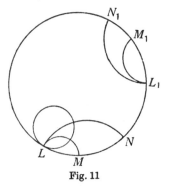

Fig. 11

For the limit-rotation obtained by inversions with respect to LM and LN, L is the only fixed point. Circles inside the non-Euclidean plane which touch the orthogonal circle at L are transformed into themselves; these circles are called *oricycles*.

47. Every ordinary circle which lies wholly within the non-Euclidean plane is also a non-Euclidean circle, and it is easy to find its non-Euclidean centre. Similarly every circle which touches the orthogonal circle is an oricycle, and every circular arc whose end-points lie on the orthogonal circle is a hypercycle. The curves of these three types are the only curves of the non-Euclidean plane which have a constant non-zero curvature. By § 44 the curvature of the oricycle is (disregarding sign) equal to 2, while that of a hypercycle is less than 2, and of a non-Euclidean circle greater than 2. If k denotes the curvature and r the non-Euclidean radius of one of these circles, we obtain the relation

$$k = 2\,\frac{e^{2r} + e^{-2r}}{e^{2r} - e^{-2r}} = 2 \coth 2r.* \qquad \ldots\ldots(47{\cdot}1)$$

48. Parallel curves.

Consider the aggregate of non-Euclidean circles with given non-Euclidean radius r whose centres are at the points of an arbitrary set A of points of the non-Euclidean plane. These circles cover a set of points $B\,(r)$, whose frontier, if it exists, contains all points of the non-Euclidean plane which are at a distance r from A.

If we take as the set A a curve C of finite curvature, and if we let r vary while remaining less than a sufficiently small upper bound, we obtain a family of parallel curves (in the sense of the non-Euclidean metric). It can be proved in just the same way as with other similar problems of the calculus of variations† that the orthogonal trajectories of such a family of equidistant curves are non-Euclidean straight lines. Conversely, if a portion of the non-Euclidean plane is simply covered by a family of non-Euclidean straight lines, the orthogonal trajectories of the family are parallel curves in the sense defined above. Thus the simplest examples of families of parallel curves are the following:

    (*a*)   non-Euclidean circles with a common centre;

    (*b*)   oricycles which touch the orthogonal circle at a common point;

    (*c*)   hypercycles having the same end-points M_1, M_2 (see Fig. 10).

* For $r < \dfrac{\pi}{2}$ we have

$$k = \frac{1}{r} + \frac{4}{3}r - \frac{16}{45}r^3 + \ldots.$$

† See Frank and v. Mises. *Die Differential- und Integralgleichungen der Mechanik und Physik.* Vol. I, ch. 5.

CHAPTER III

ELEMENTARY TRANSFORMATIONS

49. The exponential function.

The function

$$w = e^z \qquad \qquad \ldots\ldots(49\text{·}1)$$

gives rise to two important special transformations. On introducing rectangular coordinates x, y in the z-plane and polar coordinates ρ, ϕ in the w-plane, i.e. writing $z = x + iy$ and $w = \rho e^{i\phi}$, we replace (49·1) by the two equations

$$\rho = e^x, \quad \phi = y.$$

A horizontal strip of the z-plane bounded by the lines $y = y_1$ and $y = y_2$, where $|y_1 - y_2| < 2\pi$, is transformed into a wedge-shaped region of the

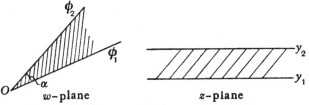

Fig. 12

w-plane, the angle of the wedge being $\alpha = |\phi_2 - \phi_1| = |y_2 - y_1|$. The representation is conformal throughout the interior of these regions, since the derivative of e^z is never zero.

As a special case, if $|y_2 - y_1| = \pi$ (e.g., $y_1 = 0$, $y_2 = \pi$), the wedge becomes a half-plane.

The restriction on the width of the strip, namely that $|y_1 - y_2| < 2\pi$, may be dropped. If, for example, $y_1 = 0$ and $y_2 = 2\pi$, the strip is repre-

Fig. 13

sented on the w-plane cut along the positive real axis; and if $|y_1 - y_2| > 2\pi$ the wedge obtained covers part of the w-plane multiply.

The cases when $|y_1 - y_2|$ is an integral multiple of 2π are of particular importance; for the strip is then transformed into a Riemann surface with an algebraic branch-point, and cut along one sheet.

50. We obtain the second special transformation furnished by (49·1) by considering an arbitrary vertical strip bounded by the lines $x = x_1$, $x = x_2$. This strip is represented on a Riemann surface which covers the annular region $\rho_1 < |w| < \rho_2$ an infinite number of times; this Riemann surface may be called the (simply-connected) *overlying surface* (*Überlagerungsfläche*) of the annular region. If now, keeping x_2 constant, we let x_1 tend to $-\infty$,

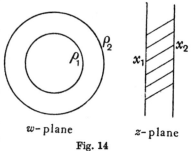

w-plane   *z*-plane

Fig. 14

the strip $x_1 < x < x_2$ becomes in the limit the half-plane $\Re(z) < x_2$, and instead of the overlying surface of an annular region we obtain a Riemann surface which covers the circle $|w| < \rho_2$ except at the point $w = 0$ where it has a logarithmic branch-point.

51. The function inverse to (49·1), namely

$$w = \log z, \qquad \qquad \ldots\ldots(51\cdot1)$$

gives, of course, on interchanging the z-plane and w-plane, precisely the same conformal transformations as (49·1).

A number of problems of conformal representation, important in the proofs of general theorems, can be solved by combining the above transformations with the Möbius' transformations discussed in Chapter I.

52. Representation of a rectilinear strip on a circle.

The strip
$$|\Re(w)| < h$$
can be represented conformally on the interior of the circle $|z| < 1$ in such a way that $w = 0$ corresponds to $z = 0$ and parallel directions through these two points also correspond. For, by § 49, the strip can be transformed into a half-plane, and then, by § 12, the half-plane into the circle. In this way we obtain, taking the half-plane to lie in the complex u-plane,

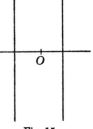

Fig. 15

$$e^{\frac{iw\pi}{2h}} = u = \frac{1 + iz}{1 - iz},$$

and hence

$$w = \frac{2h}{i\pi} \log \frac{1 + iz}{1 - iz}.$$ ......(52·1)

This gives

$$\frac{dw}{dz} = \frac{4h}{\pi(1 + z^2)},$$ ......(52·2)

which is positive when $z = 0$, as the conditions of the problem required.

If the point z describes the unit-circle $z = e^{i\theta}$, then

$$\frac{dw}{d\theta} = \frac{2hi}{\pi \cos \theta},$$

a pure imaginary as the circumstances require. If the point z describes the circle $|z| = r < 1$, w describes a curve which lies inside a finite circle and whose form is easily determined; for this curve, by (52·1),

$$|\mathfrak{F}(w)| \leqslant \frac{2h}{\pi} \log \frac{1 + r}{1 - r}.$$ ......(52·3)

The inequality (52·3) has many applications.

53. An arbitrary wedge, whose angle we denote by $\pi\alpha$, and whose vertex we suppose to be at $w = 0$, can be transformed into a half-plane. For, on introducing polar coordinates

$$z = re^{i\theta}, \quad w = \rho e^{i\phi}$$ ......(53·1)

in the relation

$$w = z^\alpha = e^{\alpha \log z},$$ ......(53·2)

this relation takes the form

$$\rho = r^\alpha, \quad \phi = \alpha\theta;$$ ......(53·3)

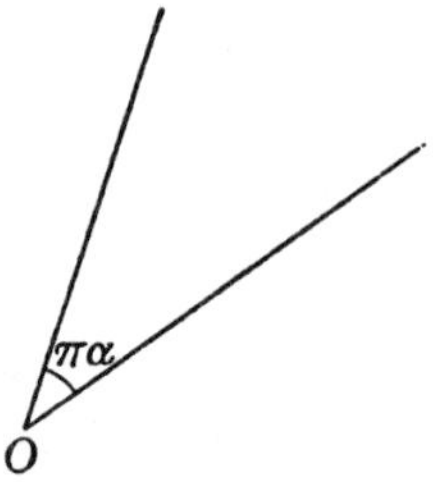

hence (53·2) effects the desired transformation *. Hence, by combination with a Möbius'

Fig. 16

transformation, a wedge can be transformed into the interior of a circle. The relations (53·3) show that the representation of a wedge on a half-plane (or on a circular area) is conformal at all points inside or on the boundary of the wedge except at its vertex $w = 0$. Two curves which intersect at $w = 0$ at an angle λ are transformed into two curves which intersect at an angle λ/α, so that at the origin corresponding angles, though no longer equal, are *proportional*; in these circumstances the representation at the origin is said to be *quasi-conformal*.

54. Representation of a circular crescent.

The area between two intersecting circular arcs or between two circles

---

* If the wedge is first transformed into a strip (§ 49), and this strip into a half-plane (§ 52), the same function (53·2) is arrived at.

having internal contact is transformed, by means of a Möbius' transformation whereby a point P common to the two circles corresponds to

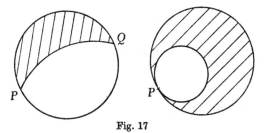

Fig. 17

the point ∞, into a wedge or a strip respectively. Hence, by the foregoing, each of these crescent-shaped areas can be represented conformally on the interior of a circle. In the same way the exterior of a circular

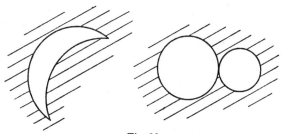

Fig. 18

crescent or the exterior of two circles having external contact can also be represented on the interior of a circle.

55. Representation of Riemann surfaces.

If, in equation (53·2), α is a positive integer n, the equation effects the representation of the unit-circle $|z| < 1$ on a portion of a Riemann surface of n sheets, overlying the circle $|w| < 1$ and having an n-fold branch-point at $w = 0$.

56. It is important to determine the function which represents the unit-circle on a Riemann surface which has the point w_0 as its branch-point but is otherwise of the kind described in § 55. We require, further, that $w = 0$ should correspond to $z = 0$ and that parallel directions drawn in the same sense through these two points should also correspond.

Suppose first that w_0 is a positive number h ($h < 1$), and displace the branch-point of the Riemann surface to the origin by means of the

Möbius' transformation (§ 35)

$$w = \frac{h-u}{1-hu}. \qquad \ldots\ldots(56\cdot1)$$

By § 55 the equation $u = t^n$ represents this new Riemann surface on the simple (*schlicht*)* circle $|t| < 1$, and in this way $t = h^{\frac{1}{n}}$ corresponds to $w = 0$. Hence the desired transformation is

$$t = \frac{h^{\frac{1}{n}} - z}{1 - h^{\frac{1}{n}} z},$$

i.e., in terms of the original variables,

$$w = h \frac{(1 - h^{\frac{1}{n}} z)^n - (1 - h^{-\frac{1}{n}} z)^n}{(1 - h^{\frac{1}{n}} z)^n - h^2 (1 - h^{-\frac{1}{n}} z)^n}. \qquad \ldots\ldots(56\cdot2)$$

The binomial theorem shows that

$$\left(\frac{dw}{dz}\right)_{z=0} = \frac{nh (h^{-\frac{1}{n}} - h^{\frac{1}{n}})}{1 - h^2} > 0. \qquad \ldots\ldots(56\cdot3)$$

In the general case when $w_0 = he^{i\theta}$ the required transformation is

$$w = h \frac{(1 - h^{\frac{1}{n}} e^{-i\theta} z)^n - (1 - h^{-\frac{1}{n}} e^{-i\theta} z)^n}{(1 - h^{\frac{1}{n}} e^{-i\theta} z)^n - h^2 (1 - h^{-\frac{1}{n}} e^{-i\theta} z)^n} e^{i\theta}. \qquad \ldots\ldots(56\cdot4,$$

57. The case in which the Riemann surface has a logarithmic branch-point at $w_0 = h$ may be treated in a similar manner. We first, by means of (56·1), transform this Riemann surface into the surface already considered in § 50; then, in virtue of the same paragraph, the transformation $u = e^t$ transforms this surface into the half-plane $\Re (t) < 0$, making $u = h$, i.e. $w = 0$, correspond to $t = \log h$. Finally the half-plane $\Re (t) < 0$ is transformed into the circle $|z| < 1$ by means of the relation

$$t = \frac{1+z}{1-z} \log h,$$

so that the function producing the desired transformation is seen to be

$$w = \frac{h - e^{\frac{1+z}{1-z} \log h}}{1 - h e^{\frac{1+z}{1-z} \log h}} = h \frac{1 - e^{\frac{2z}{1-z} \log h}}{1 - h^2 e^{\frac{2z}{1-z} \log h}}. \qquad \ldots\ldots(57\cdot1)$$

From this equation we obtain at once

$$\left(\frac{dw}{dz}\right)_{z=0} = \frac{-2h \log h}{1 - h^2} > 0. \qquad \ldots\ldots(57\cdot2)$$

---

* A Riemann surface is simple (*schlicht*) if no two points of the surface have the same coordinate u.

58. If n is made to increase indefinitely, the Riemann surface dealt with in § 56 becomes, in the limit, the Riemann surface of § 57, having a logarithmic branch-point. It is therefore to be expected that, if $\phi_n(z)$ denotes the right-hand member of (56·2), and $\psi(z)$ that of (57·1), we shall have

$$\lim_{n \to \infty} \phi_n(z) = \psi(z). \qquad \ldots\ldots(58\cdot1)$$

The truth of (58·1) can in fact be deduced from a general theorem; but the equation can also be verified directly, as follows. In

$$\frac{(h^{\frac{1}{n}} - z)^n}{(1 - h^{\frac{1}{n}} z)^n} = e^{n\,\{\log\,(h^{\frac{1}{n}} - z) - \log\,(1 - h^{\frac{1}{n}} z)\}} \qquad \ldots\ldots(58\cdot2)$$

put $h^{\frac{1}{n}} = 1 - \epsilon_n$, so that $\lim_{n \to \infty} \epsilon_n = 0$; the exponent of e in (58·2) can now be written as

$$\log h \left\{ \frac{\log\,(1 - z - \epsilon_n) - \log\,(1 - z + \epsilon_n z)}{\log\,(1 - \epsilon_n)} \right\},$$

and when ϵ_n tends to zero this tends to the limit

$$\frac{1 + z}{1 - z} \log h.$$

Equation (58·1) now follows at once. It can be shown in the same way that the limiting form of (56·3) is (57·2).

59. It can also be shown that, for all values of n,

$$\phi'_n(0) > \phi'_{n+1}(0), \qquad \ldots\ldots(59\cdot1)$$

and this inequality may perhaps rest upon some deeper, as yet unremarked property of the transformations. To prove (59·1) write $\log h^{\frac{1}{n}} = -\lambda_n$, so that $\lambda_{n+1} < \lambda_n$; but, by (56·3),

$$\phi'_n(0) = \left(\frac{-2h \log h}{1 - h^2} \right) \left(\frac{e^{\lambda_n} - e^{-\lambda_n}}{2\lambda_n} \right),$$

and the function

$$\frac{e^\lambda - e^{-\lambda}}{2\lambda} = 1 + \frac{\lambda^2}{3!} + \frac{\lambda^4}{5!} + \ldots$$

steadily decreases as λ decreases.

60. Representation of the exterior of an ellipse.

We start with the problem of representing the w-plane, cut along the finite straight line $-\alpha < w < \alpha$, where α is an arbitrary positive number, on the *exterior* $|z| > 1$ of the unit-circle, so that the points $w = \infty$ and $z = \infty$ correspond.

The Möbius' transformation

$$u = \frac{w - \alpha}{w + \alpha}, \qquad w = \alpha \frac{1 + u}{1 - u}$$

transforms the cut w-plane into the u-plane cut along the negative real axis, and the further transformation $u = t^2$ transforms this into the half-plane $\Re(t) > 0$. To the point $w = \infty$ correspond the points $u = 1$ and $t = 1$, so that the required transformation is obtained by writing

$$z = \frac{1+t}{1-t}.$$

Thus finally

$$w = \frac{\alpha}{2}\left(z + \frac{1}{z}\right); \qquad \qquad ......(60\cdot1)$$

the relation $(60\cdot1)$ is very remarkable in that it represents the cut w-plane not only on the *exterior* but also on the *interior* of the unit-circle.

61. If in $(60\cdot1)$ we write

$$z = re^{i\theta},$$

where (for example) $r > 1$, we obtain

$$w = \frac{\alpha}{2}\left\{\left(r + \frac{1}{r}\right)\cos\theta + i\left(r - \frac{1}{r}\right)\sin\theta\right\}; \qquad ......(61\cdot1)$$

hence to the circle $|z| = r$ there corresponds in the w-plane an ellipse with semi-axes

$$a = \frac{\alpha}{2}\left(r + \frac{1}{r}\right), \quad b = \frac{\alpha}{2}\left(r - \frac{1}{r}\right). \qquad ......(61\cdot2)$$

Conversely, if a, b are given, $(61\cdot2)$ determines α, r :

$$\alpha = \sqrt{(a^2 - b^2)}, \quad r = \sqrt{\left(\frac{a+b}{a-b}\right)}. \qquad ......(61\cdot3)$$

Thus equation $(60\cdot1)$ transforms the exterior of the circle $|z| > r > 1$ $\left(\text{or the interior of the circle } |z| < \frac{1}{r} < 1\right)$ into the *exterior* of the ellipse $(61\cdot1)$.

The representation of the *interior* of an ellipse on the interior of the unit-circle cannot, on the other hand, be obtained by means of the elementary transformations so far employed. But it is to be noticed that the function $(60\cdot1)$ represents the upper half of the ellipse $(61\cdot1)$ on the upper half of an annular region cut along the real axis; this last area, however, and therefore the semi-ellipse also, is easily (by a method similar to that of § 50) transformed into a rectangle. The details of this calculation, which leads to trigonometric functions, are left to the reader.

62. Representation of an arbitrary simply-connected domain on a bounded domain.

By a domain we understand an open connex (*zusammenhängend*) set of points of the complex plane; thus an open set of points is a domain if and only if every two points of the set can be joined by a continuous curve all of whose points belong to the set. The *frontier* S of a domain T is defined as the set of limiting points of T which are not also points of T. The set $S + T$ of points of a domain and of its frontier is called a closed domain and will be denoted by $\overline{T}$. In conformity with the conventions already made (see Chapter I, § 9), the point ∞ is to be treated like any other point and may, in particular, be an interior point of a domain T.

If a one-one transformation is continuous at every point of a domain T, it transforms T into another domain T'. If T is not identical with the whole complex plane, its frontier contains at least one point, and hence we can, by means of a Möbius' transformation, represent T conformally on a domain T^* which does not contain the point $z = \infty$.

63. The classification of domains according to their connectivity is important. A domain T, which may have the point ∞ as an interior point, is said to be m-ply connected if its frontier S consists of m distinct continua. The degree of connectivity is, of course, a topological invariant, that is to say, it is not altered by any continuous one-one transformation.

Our main concern here is with *simply-connected* domains, whose frontier consists of a single continuum. The property expressed by the words "simply-connected" can be specified in many other ways. It can be proved by topological methods that all these specifying properties are equivalent to one another. A few examples of such properties are:

(*a*) If the domain T^* does not contain the point ∞, then T^* is simply-connected if and only if the interior of every polygon, whose frontier belongs to T^*, consists entirely of points of T^*.

(*b*) A domain T is simply-connected if and only if every curve γ, which joins two points of the frontier S and lies within T, divides T into at least two domains.

(*c*) The same is true if every closed curve within T can be reduced to a point by continuous deformation in T. (In the course of the deformation the curve may possibly have to be taken through the point ∞.)

(*d*) A domain T is simply-connected if it can be represented on the interior of a circle by means of a continuous one-one transformation.

(*e*) The *monodromy theorem* may be regarded as giving a specifying property for a simply-connected domain T^* which does not contain the point ∞. This characterisation is especially important in the theory of functions. The theorem states:

A domain T^* which does not contain the point ∞ is simply-connected if, whenever $f(z)$ is an analytic function which can be continued along every curve in T^*, $f(z)$ is a single-valued function(11).

By this is meant: If an arbitrary functional element is assigned to a point of T^*, then analytic continuation of this element along paths entirely within T^* must *either* lead to the same functional value at any point of T^*, by whatever path that point is reached, *or* T^* must contain a singular point of the function obtained by continuation.

Every circle, for example the unit-circle $|z| < 1$, is, by this definition, a simply-connected domain. *Koebe*, in his lectures, proves this as follows:

A function $f(z)$, which can be continued along every curve within the circle, must be expressible in the neighbourhood of $z = 0$ by a power-series

$$f(z) = a_0 + a_1 z + a_2 z^2 + \dots. \qquad \dots\dots(63\cdot1)$$

The radius of convergence of this series is at least unity, for if it were less than unity analytic continuation of $f(z)$ along every radius within $|z| < 1$ would not be possible. It follows that, for all analytic continuations within $|z| < 1$, the value of the function $f(z)$ may be calculated from the relation $(63\cdot1)$, and the function is seen to be single-valued.

From this it follows that a domain is simply-connected if it can be put into one-one correspondence with the unit-circle $|z| < 1$ by means of a conformal transformation. Thus, for instance, the half-plane (§ 12), the semi-circle (§ 54), and hence also the quarter-circle (§ 53), are simply-connected. It can now be shown that the square is simply-connected, for it can be regarded as the sum of two quarter-circles; every common point P of the two quarter-circles can be joined to the middle point O of the square by means of a segment which lies entirely in $ABCM$ and entirely in $ANCD$ and the monodromy theorem applies.

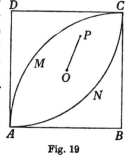

Fig. 19

64. Now let T be a simply-connected domain of the complex z-plane, such that the frontier of T contains at least two points A_1, B_1, and let z_0 be an interior point of T, other than the point ∞. Consider the circular arc (which may be a finite or infinite straight line) which joins

A_1 to B_1 and has z_0 as an interior point, and denote by A the first frontier-point of T which is met in describing this arc from z_0 to A_1, and by B the corresponding point of the arc from z_0 to B_1. Thus all interior points of the arc Az_0B are interior points of T, and its end-points are frontier-points of T. If we now apply to the z-plane the Möbius' transformation whereby the points A, z_0, B correspond to the points 0, 1, ∞ in the u-plane, the domain T is transformed into a simply-connected domain T_1, which is such that $u = 0$ and $u = \infty$ are frontier-points of T_1 but all other points of the positive real axis are interior points of T_1.

By means of the transformation $u = v^4$ we now obtain four different domains in the v-plane corresponding to T_1; let T_2 denote that one of them in which the straight line $v > 0$ corresponds to the straight line $u > 0$*. We shall prove that T_2 lies in the half-plane $\Re(v) > 0$, and is therefore transformed by

$$w = \frac{v-1}{v+1} \qquad \qquad ......(64\cdot1)$$

into a domain T' which lies inside the circle $|w| < 1$. For if T_2 were to contain a point v_2 for which $\Re(v) \leqslant 0$, we could draw in T_2 a curve joining v_2 and $v_1 = 1$; let P be the first point of intersection (counting from v_1) of this curve with the imaginary v-axis, and Q the last intersection of the arc from v_1 to P with the real axis, so that the arc PQ, which we will denote by γ_v, lies either in the first or in the fourth quadrant of the v-plane. The relation $u = v^4$ transforms the arc γ_v into a curve γ_u all of whose points belong to T_1, and whose end-points lie on the real axis $u > 0$; by adding to γ_u a portion of the axis $u > 0$ we therefore obtain a closed curve which lies wholly in T_1 and surrounds the point $u = 0$. But no such curve can exist, since T_1 is simply-connected and does not contain either of the points $u = 0$, $u = \infty$.

Thus: *any simply-connected domain whose frontier contains at least two distinct points can, by simple transformations, be conformally represented on a domain which lies entirely inside the unit-circle.*

A multiply-connected domain can be treated in precisely the same way provided that its frontier contains at least one continuum of more than one point.

65. A theorem which, though in appearance but little different from that of § 64, is, on account of the value of the constant involved, of the greatest importance in the general theory, is due to *Koebe*[12].

---

* Each of the four domains corresponding to T_1, in particular T_2, is simple (*schlicht*) (cf. § 56, footnote).

Let T be a simple simply-connected domain of the w-plane not containing the point $w = \infty$; suppose further that all points of the circle $|w| < 1$ are points of T, but that $w = 1$ is a frontier-point of T. Any such domain T is transformed by the same function

$$w = \frac{4(1 + \sqrt{2})^2 z}{\{1 + (1 + \sqrt{2})^2 z\}^2} \qquad \ldots\ldots (65\cdot1)$$

into a simple simply-connected domain T' which lies inside the unit-circle $|z| < 1$ but always contains the fixed circle $|z| < (1 + \sqrt{2})^{-4}$.

Consider in the w-plane the two-sheeted Riemann surface which has branch-points at $w = 1$ and $w = \infty$; the function

$$1 - w = u^2 \qquad \ldots\ldots (65\cdot2)$$

transforms this surface into the simple u-plane, making the two points $u = \pm 1$ correspond to $w = 0$, and transforms any curve drawn in T and starting from $w = 0$ into *two* curves in the u-plane. The aggregate of all curves in the u-plane which correspond to curves in T and which start from $u = +1$ fills a simply-connected domain which we will denote by T_1.

We shall prove that, if u_0 is any complex number, the domain T_1 cannot contain *both* the points $\pm u_0$. For if it did, these two points could be joined by a curve γ_u lying inside T_1, and to γ_u there would correspond in the w-plane a *closed* curve γ_w lying inside T, some point w_0 of γ_w corresponding to the end-points $\pm u_0$ of γ_u. Since T is simply-connected we can now, by continuous deformation, and keeping the curve always inside T, reduce γ_w to an arbitrarily small curve passing through w_0. But as, in this deformation, the end-points of γ_u remain fixed, we have clearly arrived at a contradiction.

66. Equation $(65\cdot2)$ transforms the circle $|w| = 1$ into a bicircular quartic curve whose equation in polar coordinates is

$$\rho^2 = 2 \cos 2\theta. \qquad \ldots\ldots (66\cdot1)$$

This curve is symmetrical with respect to the origin; and, of the two domains bounded by its loops, one must lie entirely inside T_1, and hence, by § 65, the other must lie entirely outside T_1.

The function

$$t = \frac{1 - u}{1 + u} \qquad \ldots\ldots (66\cdot2)$$

transforms the curve $(66\cdot1)$ into the curve C of Fig. 21, having a double point at $t = 1$; and it transforms T_1 into a domain T_2 which lies entirely inside the curve C, since the exterior of C corresponds to that loop of the curve $(66\cdot1)$ which contains no point of T_1. It therefore follows that the inner loop of C lies entirely inside T_2. But (65·2) and (66·2) give

$$t = \frac{1 - \sqrt{(1-w)}}{1 + \sqrt{(1-w)}} = \frac{\{1 - \sqrt{(1-w)}\}^2}{w}; \qquad \ldots\ldots(66\cdot3)$$

and from this we obtain at once that, if $|w| = 1$, $|t| \leqslant (1 + \sqrt{2})^2$. Consequently T_2 lies entirely inside this last circle; also, since the curve C is its own inverse with respect to the unit-circle $|t| = 1$, we see further that T_2 must contain the whole circle $|t| = (1 + \sqrt{2})^{-2}$ in its interior.

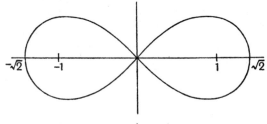

u-plane

Fig. 20

The theorem enunciated at the beginning of § 65 is now obtained on using (66·3) to express w as a function of t and then writing

$$t = (1 + \sqrt{2})^2 z. \qquad \ldots\ldots(66\cdot4)$$

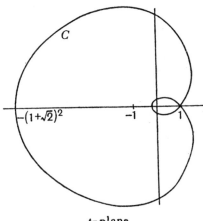

t-plane

Fig. 21

Since we supposed that T does not contain the point $w = \infty$, T_1 does not contain the point $u = \infty$; hence $t = -1$ lies outside T_2, and, finally, $z = -(1 + \sqrt{2})^{-2}$ lies outside T'.

Remark. Let T^* be a simply-connected domain contained in T and containing the point $w = 0$, and denote the frontier of T^* by γ_w. The transformation (65·2) gives two distinct domains in the u-plane corresponding to T^*. One of these contains the point $u = +1$ and the other the point $u = -1$, and their boundaries may be denoted by γ_u' and γ_u'' respectively. If u' is a point of γ_u', the relation $u'' = -u'$ transforms it into a point u'' of γ_u''. The relation (66·2) transforms these two domains into distinct simply-connected domains in the t-plane. One of these, say T_t^*, contains the point $t = 0$ and consists of the points lying within a certain continuum γ_t'; the other consists of the points lying outside a continuum γ_t''. Since the domains have no points in common, γ_t'' surrounds γ_t'. The relation $t'' = \dfrac{1}{t'}$ transforms γ_t' and γ_t'' into one another.

CHAPTER IV

SCHWARZ'S LEMMA

67. Schwarz's Theorem.

We take as starting-point the observation that if an arbitrary analytic function $f(z)$ is regular in a closed domain $\bar{T}$, the maximum value of $|f(z)|$ in $\bar{T}$ is attained at one point at least of the frontier S of $\bar{T}$. The familiar proof of this theorem depends upon the elementary fact that every neighbourhood of a regular point z_0 of a non-constant function $f(z)$ contains points z_1 such that $|f(z_1)| > |f(z_0)|$.

Consider now a function $f(z)$ satisfying the following three conditions: (i) $f(z)$ is regular in the circle $|z| < 1$; (ii) at all points of this circle $|f(z)| < 1$; (iii) $f(0) = 0$. From conditions (i) and (iii) it follows that

$$\phi(z) = f(z)/z \qquad \ldots\ldots(67\cdot1)$$

is also regular in $|z| < 1$; consequently, if z_0 is any point inside the unit-circle, and r denotes a positive number lying between $|z_0|$ and unity, there is on the perimeter of the closed circle $|z| \leqslant r$ at least one point z_1 such that $|\phi(z_1)| \geqslant |\phi(z_0)|$. But, by hypothesis, $|f(z_1)| < 1$ and $|z_1| = r$, so that we may write

$$|\phi(z_0)| \leqslant |\phi(z_1)| = \frac{|f(z_1)|}{|z_1|} < \frac{1}{r};$$

hence, since r may be taken arbitrarily near to unity,

$$|\phi(z_0)| \leqslant 1. \qquad \ldots\ldots(67\cdot2)$$

Thus for all points inside the unit-circle $|\phi(z)| \leqslant 1$; but if $\phi(z)$ is not constant, this may be replaced by $|\phi(z)| < 1$ for all interior points. For if, $\phi(z)$ not being constant, there were an interior point z_0 such that $|\phi(z_0)| = 1$, there would necessarily exist other interior points for which $|\phi(z)| > 1$, and this contradicts $(67\cdot2)$. On use of $(67\cdot1)$ there now follows:

THEOREM 1. SCHWARZ'S LEMMA. *If the analytic function $f(z)$ is regular for $|z| < 1$, if $|f(z)| < 1$ for $|z| < 1$, and if further $f(0) = 0$, then either*

$$|f(z)| < |z| \qquad \ldots\ldots(67\cdot3)$$

for $|z| < 1$ or $f(z)$ is a linear function of the form

$$f(z) = e^{i\theta} z, \qquad \ldots\ldots(67\cdot4)$$

where θ is real[13].

By (67·1) $\phi(0) = f'(0)$, and our former reasoning therefore gives at once the further theorem:

THEOREM 2. *If $f(z)$ satisfies the conditions of Schwarz's Lemma, then either $|f'(0)| < 1$ or $f(z)$ is of the form (67·4).*

68. Theorem of uniqueness for the conformal representation of simply-connected domains.

One of the simplest but most important applications of Schwarz's Lemma is the following. Let G be a simply-connected domain in the w-plane, containing, we will suppose, the point $w = 0$ in its interior; and we assume that a function $w = f(z)$ exists which represents G conformally on the circle $|z| < 1$, and is such that $f(0) = 0$ and $f'(0)$ is real and positive. We shall prove that *the function $f(z)$ satisfying these conditions is necessarily unique.*

For if there were a second function $g(t)$, distinct from $f(z)$, but satisfying the same conditions, the equation

$$g(t) = f(z) \qquad \qquad \ldots\ldots(68\cdot1)$$

would transform the circle $|t| < 1$ into the circle $|z| < 1$, making the centres $t = 0$, $z = 0$ correspond to each other. Hence *both* the functions

$$t = \phi(z), \quad z = \psi(t), \qquad \qquad \ldots\ldots(68\cdot2)$$

obtained by solving (68·1), would satisfy all the conditions of Schwarz's Lemma, and hence, by §67, both the inequalities $|t| \leqslant |z|$, $|z| \leqslant |t|$ would hold for all values of z and t satisfying (68·2). Thus $|t| = |\phi(z)| = |z|$ and, again by Schwarz's Lemma, $\phi(z)$ would be of the form $e^{i\theta}z$. But it follows from the original hypotheses that $\phi'(0) > 0$, and hence $\phi(z) \equiv z$; that is to say

$$f(z) = g(\phi(z)) = g(z),$$

so that the function $f(z)$ is unique.

69. Liouville's Theorem.

A large number of important theorems follow readily on combining Schwarz's Lemma with earlier theorems. As an example we may mention the theorem of *Liouville* that a bounded integral function is necessarily a constant.

For if $F(z)$ is an integral function such that $|F(z)| < M$ for all values of z, and we write $z = Ru$,

$$f(u) = \frac{F(z) - F(0)}{2M},$$

the function $f(u)$ satisfies the conditions of Schwarz's Lemma for any positive value of R. Hence, if $|u| < 1$, $|f(u)| < u$; i.e. if $|z| < R$,

$$|F(z) - F(0)| < \frac{2M}{R}|z|.$$

Keeping z constant and letting R tend to infinity we obtain $F(z) = F(0)$, as was asserted.

70. Invariant enunciation of Schwarz's Lemma.

In the circle $|z| < 1$ let the function $w = f(z)$ be regular and let $|f(z)| < 1$; we no longer suppose, however, that $f(0) = 0$. Denoting by z_0 any point inside the unit-circle, transform the unit-circles in the z-plane and in the w-plane into themselves by means of the Möbius' transformations (cf. § 35)

$$t = \frac{z - z_0}{1 - \bar{z}_0 z}, \qquad \omega = \frac{w - f(z_0)}{1 - \overline{f}(z_0) w} = \frac{f(z) - f(z_0)}{1 - \overline{f}(z_0) f(z)} \dots \dots (70 \cdot 1)$$

The function $\omega = \omega(t)$ obtained by elimination of z from equations $(70 \cdot 1)$ satisfies all three conditions of Schwarz's Lemma, and consequently

$$|\omega(t)| \leqslant |t|. \qquad \dots \dots (70 \cdot 2)$$

This inequality can be expressed by the statement that the non-Euclidean distance $D(0, \omega(t))$ of the two points $\omega(0) = 0$ and $\omega(t)$ is not greater than the non-Euclidean distance of the points 0, t, so that, remembering that, as was proved in Chapter II, non-Euclidean distances are invariant for the transformations $(70 \cdot 1)$, we obtain the following theorem, which includes Schwarz's Lemma as a special case and which was first stated by *G. Pick*[14].

THEOREM 3. *Let $f(z)$ be an analytic function which is regular and such that $|f(z)| < 1$ in the circle $|z| < 1$, and let z_1, z_2 denote any two points inside the unit-circle; then either*

$$D(f(z_1), f(z_2)) < D(z_1, z_2) \qquad \dots \dots (70 \cdot 3)$$

for all such values of z_1 and z_2, or

$$D(f(z_1), f(z_2)) = D(z_1, z_2) \qquad \dots \dots (70 \cdot 4)$$

for all such values of z_1 and z_2. Further, if $(70 \cdot 4)$ holds, the function $w = f(z)$ is necessarily a Möbius' transformation which transforms the unit-circle into itself.

71. Comparing $(70 \cdot 2)$ with $(70 \cdot 1)$ we obtain

$$\left| \frac{f(z) - f(z_0)}{1 - \overline{f}(z_0) f(z)} \right| \leqq \frac{|z - z_0|}{|1 - \bar{z}_0 z|} \quad \text{or} \quad \left| \frac{f(z) - f(z_0)}{z - z_0} \right| \leqq \left| \frac{1 - \overline{f}(z_0) f(z)}{1 - \bar{z}_0 z} \right|;$$

and if we let z converge towards z_0 this gives

$$|f'(z_0)| \leqslant \frac{1 - |f(z_0)|^2}{1 - |z_0|^2}.$$

Thus, since z_0 is arbitrary, we now have the theorem:

THEOREM 4. *If $f(z)$ satisfies the conditions of Theorem 3, then either*

$$|f'(z)| < \frac{1 - |f(z)|^2}{1 - |z|^2} \qquad \ldots\ldots(71\text{·}1)$$

for all points z inside the unit-circle, or

$$|f'(z)| = \frac{1 - |f(z)|^2}{1 - |z|^2} \qquad \ldots\ldots(71\text{·}2)$$

for all points z inside the unit-circle; further $(71\text{·}2)$ holds if and only if $(70\text{·}4)$ holds.

From $(71\text{·}1)$ and $(71\text{·}2)$ it follows at once that, if $|z| \leqslant r < 1$,

$$|f'(z)| \leqslant \frac{1}{1 - r^2}; \qquad \ldots\ldots(71\text{·}3)$$

hence, denoting by z_1 and z_2 any two points of the closed circle $|z| \leqslant r$, and integrating along the straight line from z_1 to z_2, the relation

$$\int_{z_1}^{z_2} f'(z)\, dz = f(z_2) - f(z_1)$$

gives at once, by the Mean Value theorem,

THEOREM 5. *If the function $f(z)$ satisfies the conditions of Theorem 3, and if z_1, z_2 are any two points of the closed circle $|z| \leqslant r < 1$, then*

$$\left| \frac{f(z_1) - f(z_2)}{z_1 - z_2} \right| \leqslant \frac{1}{1 - r^2}. \qquad \ldots\ldots(71\text{·}4)$$

72. As a last application of equations $(70\text{·}1)$ and $(70\text{·}2)$ we take the following: if we solve the second of equations $(70\text{·}1)$ for $f(z)$ and take $z_0 = 0$, we obtain

$$f(z) = \frac{\omega + f(0)}{1 + \bar{f}(0)\,\omega}; \qquad \ldots\ldots(72\text{·}1)$$

but, for $z_0 = 0$, the first of equations $(70\text{·}1)$ becomes $t = z$, so that, by $(70\text{·}2)$, $|\omega| \leqslant |z|$. Also $(72\text{·}1)$ is a Möbius' transformation which transforms the circle $|\omega| \leqslant |z|$ into another easily constructed circle, and by consideration of the figure so obtained we arrive without difficulty at the theorem:

THEOREM 6. *If $f(z)$ is regular and $|f(z)| < 1$ in the circle $|z| < 1$, then*

$$|f(z)| \leqslant \frac{|z| + |f(0)|}{1 + |f(0)||z|} \qquad \ldots\ldots(72\cdot2)$$

at all points z inside the unit-circle.

Finally we deduce from (72·2), by an easy manipulation,

$$\frac{1 - |f(z)|}{1 - |z|} \geqslant \frac{1 - |f(0)|}{1 + |f(0)||z|} \geqslant \frac{1 - |f(0)|}{1 + |f(0)|}. \qquad \ldots\ldots(72\cdot3)$$

73. Theorem 3 of §70 may be regarded as a special case of a more general theorem which we will now prove. If the function $w = f(z)$ is regular for $|z| < 1$, every curve γ_z in the z-plane which lies entirely inside the unit-circle corresponds to a curve γ_w in the w-plane, which will also lie entirely inside the unit-circle provided that $|f(z)| < 1$ for the values of z considered. If the curve γ_z is rectifiable (in the ordinary sense) it has, by §43, a non-Euclidean length $L(\gamma_z)$ which is defined as the upper limit of the non-Euclidean lengths of certain inscribed curvilinear polygons; to any non-Euclidean polygon P_w inscribed in γ_w there corresponds a curvilinear polygon P_z inscribed in γ_z, and, by §70, the non-Euclidean distance between two consecutive vertices of P_w is never greater than that between the two corresponding vertices of P_z. From this it follows at once that the curve γ_w is also rectifiable and has a non-Euclidean length $L(\gamma_w)$ which cannot exceed $L(\gamma_z)$.

If now the two curves have the same non-Euclidean length, corresponding elementary arcs of γ_z and γ_w must also be equal, so that (71·2) must be satisfied at all points of γ_z. Thus, on using Theorem 4 of §71, we obtain the required theorem:

THEOREM 7. PICK'S THEOREM. *If the function $w = f(z)$ is regular and $|f(z)| < 1$ in the circle $|z| < 1$, and if $L(\gamma_z)$, $L(\gamma_w)$ denote the non-Euclidean lengths of corresponding arcs γ_z, γ_w drawn in the unit-circle, then either*

$$L(\gamma_w) < L(\gamma_z) \qquad \ldots\ldots(73\cdot1)$$

for all such arcs, or else the function $w = f(z)$ defines a non-Euclidean motion, and in the latter case

$$L(\gamma_w) = L(\gamma_z) \qquad \ldots\ldots(73\cdot2)$$

for all such arcs.

74. Functions with positive real parts.

In the circle $|z| < 1$ suppose that $f(z)$ is regular, that $\Re f(z) > 0$ and also suppose $f(0) = 1$. By the Möbius' transformation

$$u = \frac{1 + w}{1 - w}, \quad w = \frac{u - 1}{u + 1}, \qquad \ldots\ldots(74\cdot1)$$

the half-plane $\mathfrak{R}u > 0$ is transformed into the unit-circle $|w| < 1$, and $u = 1$, $w = 0$ are corresponding points. Schwarz's Lemma can therefore be applied to the function

$$w = \frac{f(z) - 1}{f(z) + 1}. \qquad \text{......(74·2)}$$

Consequently, the figure in the u-plane, which corresponds, by means of the relation $u = f(z)$, to the circle $|z| < r$ $(0 < r < 1)$ must lie inside the circle which corresponds, by means of the Möbius' transformation (74·1), to the circle $|w| < r$. If the figures are drawn, it is seen at once that, with our assumptions, $|z| < r$ leads to the relations

$$\frac{1-r}{1+r} < |f(z)| < \frac{1+r}{1-r}, \quad \frac{1-r}{1+r} < \mathfrak{R}f(z) < \frac{1+r}{1-r}, \quad \text{......(74·3)}$$

$$|\mathfrak{I}f(z)| < \frac{2r}{1-r^2}. \qquad \text{......(74·4)}$$

75. Harnack's Theorem.

It will now be shown that a theorem of *Harnack*, the fundamental importance of which has been long recognised, is an almost obvious consequence of our last result, so that Harnack's Theorem may without loss be replaced by Schwarz's Lemma.

Harnack's Theorem may be stated as follows: *If U_1, U_2, U_3, ... denote a monotone increasing sequence of harmonic functions in the circle $|z| < 1$, and if U_n tends to a finite limit at $z = 0$, then U_n tends to a limit uniformly in the circle $|z| < r < 1$.*

We write

$$u_n = U_{n+1} - U_n, \qquad \text{......(75·1)}$$

$$f_n(z) = u_n + iv_n, \qquad \text{......(75·2)}$$

where v_n is the harmonic function conjugate to u_n and vanishing at the origin, so that, by hypothesis, $\mathfrak{R}f_n(z) = u_n \geqslant 0$, and consequently $u_n(0) = f_n(0) > 0$ unless $f_n(z)$ is identically zero. The functions $f_n(z)/u_n(0)$ therefore satisfy the conditions of § 74, so that, by (74·3),

$$|f_n(z)| < u_n(0)\frac{1+r}{1-r}$$

if $|z| < r$; hence also

$$u_n < u_n(0)\frac{1+r}{1-r},$$

and Harnack's Theorem follows at once from this.

76. Functions with bounded real parts.

Suppose the function $f(z)$ to satisfy the following conditions:
(i) $f(0) = 0$; (ii) $f(z)$ is regular for $|z| < 1$; (iii) there is a constant h such that

$$|\Re f(z)| < h \qquad \qquad \ldots\ldots(76\cdot1)$$

if $|z| < 1$.

Write $f(z) = w$ and, using (52·1),

$$w = \frac{2h}{i\pi} \log \frac{1 + i\omega}{1 - i\omega}. \qquad \qquad \ldots\ldots(76\cdot2)$$

These relations determine ω as a function $\omega(z)$ of z, and $\omega(z)$ satisfies the conditions of Schwarz's Lemma. From this fact and the relation (52·3) for $|z| < r$ that

$$|\Im f(z)| \leqslant \frac{2h}{\pi} \log \frac{1 + r}{1 - r}. \qquad \qquad \ldots\ldots(76\cdot3)$$

77. Surfaces with algebraic and logarithmic branch-points.

The functions $\phi_n(z)$ and $\psi(z)$, by means of which, in § 56 and § 57, the unit-circle was represented on the circular areas having algebraic or logarithmic branch-points, satisfy all conditions of Schwarz's Lemma; therefore they must satisfy the inequalities

$$|\phi_n(z)| < |z|, \quad |\psi(z)| < |z|; \qquad \ldots\ldots(77\cdot1)$$
$$|\phi_n'(0)| < 1, \quad |\psi'(0)| < 1. \qquad \ldots\ldots(77\cdot2)$$

The inequalities (77·2) can also easily be verified directly; for in § 59 it was shown that, if $n > 2$, $|\psi'(0)| < |\phi_n'(0)| < |\phi_2'(0)|$. But, by (56·3),

$$\phi_2'(0) = \frac{2\sqrt{h}}{1 + h} < 1.$$

78. A further property of the function used in § 57,

$$\psi(u) = \frac{h - e^{\frac{1+u}{1-u}\log h}}{1 - h e^{\frac{1+u}{1-u}\log h}}, \; (h < 1), \qquad \ldots\ldots(78\cdot1)$$

may be obtained as follows. Consideration of the Möbius' transformation employed shows that, if $|u| < r < 1$,

$$\frac{1 + r}{1 - r} \log h < \Re \left(\frac{1 + u}{1 - u} \log h \right) < \frac{1 - r}{1 + r} \log h < 0, \; \ldots\ldots(78\cdot2)$$

and hence, in particular

$$\left| e^{\frac{1+u}{1-u}\log h} \right| > e^{\frac{1+r}{1-r}\log h} = \rho. \qquad \ldots\ldots(78\cdot3)$$

Consequently, by (78·1), the values assumed by $\psi(u)$ for these values of u must lie *outside* a non-Euclidean circle with non-Euclidean centre h, and from this we at once obtain, remembering that $\rho < 1$, the inequality

$$|\psi(u) - h| > \frac{h + \rho}{1 + h\rho} - h > (1 - h)\rho,$$

i.e.
$$|\psi(u) - h| > (1 - h) e^{\frac{1+r}{1-r} \log h}. \qquad \ldots\ldots(78\cdot4)$$

79. Consider now a function $w = f(z)$ having the three properties (i), (ii), and (iii) of § 67, and having also the further property that there is a real number h, $(0 < h < 1)$ such that the equation $f(z) = h$ has no solution in the circle $|z| < 1$. With these assumptions it may be shown that if from the equation

$$\psi(u) = f(z) \qquad \ldots\ldots(79\cdot1)$$

(where $\psi(u)$ denotes the same function as in $(78\cdot1)$) we obtain that solution $u = \phi(z)$ which vanishes at $z = 0$, the function $\phi(z)$ possesses all the properties (i), (ii), (iii) of § 67. Hence $|u| < r$ if $|z| < r$, and comparison of $(78\cdot4)$ with $(79\cdot1)$ yields the following theorem:

THEOREM. *If $f(z)$ is regular for $|z| < 1$, and if in this circle $|f(z)| < 1$, $f(z) \neq h$ $(0 < h < 1)$, and, finally, if $f(0) = 0$, then in any circle $|z| < r < 1$*

$$|f(z) - h| > (1 - h) e^{\frac{1+r}{1-r} \log h}. \qquad \ldots\ldots(79\cdot2)$$

80. Representation of simple domains.

Let the function

$$w = f(z) \qquad \ldots\ldots(80\cdot1)$$

represent the simple domain T, which was discussed in § 65, on the interior of the unit-circle $|z| < 1$, and suppose that $f(0) = 0$. If in $(65\cdot1)$ we replace z by $-\omega$, and if we denote $(1 + \sqrt{2})^{-2}$ by h, §§ 65 and 66 show that the function

$$\omega = \phi(z), \qquad \ldots\ldots(80\cdot2)$$

obtained by solving the equation

$$\left.\begin{array}{l} w = \dfrac{-4h\omega}{(h - \omega)^2} = f(z), \\ h = (1 + \sqrt{2})^{-2}, \end{array}\right\} \qquad \ldots\ldots(80\cdot3)$$

effects a conformal representation of the domain T' considered at the end of § 66 which lies inside the unit-circle $|\omega| < 1$ and does not contain the point $\omega = h$. Since $\phi(0) = 0$ it follows from Schwarz's Lemma that $|\omega| \leqslant r$ if $|z| \leqslant r$, and also, by the Theorem of § 79, that

$$|\omega - h| > (1 - h) e^{\frac{1+r}{1-r} \log h}. \qquad \ldots\ldots(80\cdot4)$$

But, from the second of equations (80·3),

$$1 - h = 2\sqrt{h},$$

and hence, from the first equation of (80·3), using the inequalities $|\omega| \leqslant r$ and (80·4), if $|z| \leqslant r$,

$$|f(z)| < re^{-2\frac{1+r}{1-r}\log h}. \qquad \ldots\ldots(80\cdot5)$$

Observing that $1 + \sqrt{2} = 2\cdot41 \ldots < 2\cdot7 \ldots < e$, so that

$$-2\log h = 4\log(1 + \sqrt{2}) < 4,$$

we see that (80·5) includes the inequality

$$|f(z)| < re^{4\frac{1+r}{1-r}}. \qquad \ldots\ldots(80\cdot6)$$

Thus we have proved that, in the conformal representation (80·1), the domain corresponding to the circle $|z| \leqslant r < 1$ is bounded and lies wholly within the fixed circle $|w| \leqslant re^{4\frac{1+r}{1-r}}$.

On the other hand the domain T contains, by hypothesis (see § 65), the circle $|w| < 1$ in its interior, and hence, by Schwarz's Lemma applied to the function $z = \phi(w)$ inverse to $f(z)$, the figure corresponding to $|z| \leqslant r$ must contain the circle $|w| \leqslant r$ in its interior. If we now apply an arbitrary magnification, we obtain the following general theorem :

THEOREM. *Let T be a simple domain of the w-plane containing $w = 0$ in its interior but not containing $w = \infty$; let a be the distance of the point $w = 0$ from the frontier of T, and let $f(z)$ denote a function which represents T conformally on the circle $|z| < 1$, making $w = 0$ correspond to $z = 0$. Then, for any value of z inside the circle $|z| < 1$*,

$$a \leqslant \frac{|f(z)|}{|z|} < ae^{4\frac{1+|z|}{1-|z|}}. \qquad \ldots\ldots(80\cdot7)$$

The formulae already established can also be used to obtain limits for the difference quotient $\{f(z_1) - f(z_2)\}/(z_1 - z_2)$ of the function $f(z)$. For example, by (80·7), the function $f(ru)/are^{4\frac{1+r}{1-r}}$ has modulus less than unity inside the circle $|u| < 1$. Hence by Theorem 5 of § 71 inside the circle $|u| < r$, i.e. inside the circle $|z| < r^2$,

$$\left|\frac{f(z_1) - f(z_2)}{z_1 - z_2}\right| = \left|\frac{f(ru_1) - f(ru_2)}{r(u_1 - u_2)}\right| \leqq \frac{ae^{4\frac{1+r}{1-r}}}{1 - r^2},$$
$$|z_1| < r^2, \quad |z_2| < r^2.$$

81. By letting $|z|$ tend to zero in (80·7) we see that the function $f(z)$ of § 80 satisfies the inequalities $a \leqq |f'(0)| \leqq ae^4$. It is known by Schwarz's

Lemma that a is the true lower bound for $|f'(0)|$; we shall now deter-
mine the true upper bound for this same number, using a very ingenious
method due to *Erhard Schmidt*. It is published here for the first time[15].

The considerations which follow form the basis of the proof :

(*a*) Suppose that the function $f(z)$ is not a constant and that it is
regular within and on the frontier of a simply-connected domain D, whose
frontier is a regular curve γ. Then, if $f(z) = u + iv$, where u and v are real,

$$\int_{\gamma} u\,dv > 0. \qquad \qquad \dots\dots(81\cdot1)$$

For, if account is taken of the Cauchy-Riemann equations, Green's
Theorem shows that

$$\int_{\gamma} u\,dv = \iint_{D} (u_x{}^2 + u_y{}^2)\,dx\,dy$$

(*b*) The inequality (81·1) can be extended to multiply-connected
domains provided $f(z)$ is regular and single-valued in the domain con-
sidered. Let γ' and γ'' be two closed curves such that γ' lies within γ''
and itself surrounds $z = 0$. Then the function $\log z$ is regular in the annular domain between γ' and γ'' but it is not single-valued. If however
$z = \rho e^{i\psi}$ and $\log z = \log \rho + i\psi$, then $\log \rho$ and $d\psi$ are single-valued functions in the annular region, and hence

$$\int_{\gamma''} \log \rho\,d\psi \geqslant \int_{\gamma'} \log \rho\,d\psi. \qquad \dots\dots(81\cdot2)$$

(*c*) If the curves γ' and γ'' are such that the relation $z'' = \dfrac{1}{z}$ transforms
γ' into γ'', it is easy to see that

$$\int_{\gamma''} \log \rho\,d\psi + \int_{\gamma'} \log \rho\,d\psi = 0. \qquad \dots\dots(81\cdot3)$$

From (81·2) and (81·3) it is seen that in this special case

$$\int_{\gamma'} \log \rho\,d\psi \leqslant 0. \qquad \dots\dots(81\cdot4)$$

(*d*) Let $F(z) = z\phi(z)$, where $\phi(z)$ is regular and differs from zero
throughout the closed circle $|z| \leqslant 1$. We write

$$F(e^{i\theta}) = \rho(\theta)\,e^{i\psi(\theta)}. \qquad \dots\dots(81\cdot5)$$

The hypothesis shows that the function $\log \phi(z)$ is regular for $|z| \leqslant 1$,
and, if $|z| = 1$, then $\log \phi(e^{i\theta}) = \log (F(e^{i\theta})\,e^{-i\theta}) = \log \rho + i(\psi - \theta)$. By
(81·1)

$$\int_{\kappa} \log \rho\,d(\psi - \theta) \geqslant 0, \qquad \dots\dots(81\cdot6)$$

where κ is the circle $|z| = 1$.

On the other hand, by the Mean Value Theorem,

$$\log|\phi(0)| = \Re \log \phi(0) = \frac{1}{2\pi} \int_0^{2\pi} \log \rho \, d\theta, \quad \ldots\ldots(81\cdot7)$$

and, further, $\phi(0) = F'(0)$. Comparison of $(81\cdot6)$ and $(81\cdot7)$ shows that

$$\log|F'(0)| \leqslant \frac{1}{2\pi} \int_\kappa \log \rho \, d\psi. \quad \ldots\ldots(81\cdot8)$$

82. Let $f(z)$ be the function considered at the end of §80. If $r < 1$ we write

$$g(z) = \frac{1}{a} f(rz) \quad \ldots\ldots(82\cdot1)$$

and

$$F(z) = \frac{1 - \sqrt{1 - g(z)}}{1 + \sqrt{1 - g(z)}} = \frac{g(z)}{\{1 + \sqrt{1 - g(z)}\}^2} . \quad \ldots\ldots(82\cdot2)$$

By $(66\cdot3)$, the function $F(z)$ transforms $|z| \leqslant 1$ conformally into a simple domain T_2^* which lies within the domain T_2 of §66. Thus $\dfrac{F(z)}{z}$ does not vanish in the unit-circle. If therefore

$$F(e^{i\theta}) = \rho e^{i\psi},$$

the inequality $(81\cdot8)$ holds. By $(82\cdot2)$ and $(82\cdot1)$

$$F'(0) = \frac{g'(0)}{4} = \frac{r}{4a} f'(0).$$

Thus

$$\log\left|\frac{rf'(0)}{4a}\right| \leqslant \frac{1}{2\pi} \int_\kappa \log \rho \cdot \psi'(\theta) \, d\theta = \frac{1}{2\pi} \int_{\gamma'} \log \rho \, d\psi.$$
$$\ldots\ldots(82\cdot3)$$

Reference to the last part of §66 now shows that γ' has all the properties assumed for this curve in (b) and (c) in §81, so that $(81\cdot4)$ holds. From this and $(82\cdot3)$ it follows that $|f'(0)| \leqslant 4a/r$, and, since this holds for all $r < 1$,

$$|f'(0)| \leqslant 4a. \quad \ldots\ldots(82\cdot4)$$

This upper bound cannot be improved upon, for the function

$$w = f(z) = a \frac{4z}{(1+z)^2}$$

transforms the circle $|z| < 1$ into a simple domain whose frontier is at a distance a from $w = 0$.

83. Representation upon one another of domains containing circular areas.

Let R_w and R_z be two simply-connected domains which lie entirely inside the circles $|w| < 1$ and $|z| < 1$, but contain the circles $|w| < h < 1$ and $|z| < h$ in their interiors. Suppose further that there is a function $w(z)$, which of course need only be defined inside R_z, such that $w(z)$ represents R_z on R_w, making $z = 0$ correspond to $w = 0$, and making parallel directions through these two points correspond; i.e. $w(0) = 0$, and $w'(0)$ is real and positive. If then $0 < r < 1$ we shall show that, in the circle $|z| \leqslant hr$,

$$|w(z) - z| \leqslant m(h, r), \qquad \text{......(83·1)}$$

where the function $m(h, r)$ is independent of the configurations of the parts of R_w and R_z which lie outside the circles $|w| < h$ and $|z| < h$, and where, further,

$$\lim_{h \to 1} m(h, r) = 0. \qquad \text{......(83·2)}$$

In the first place, by Schwarz's Lemma,

$$\left| \frac{w(z)}{z} \right| \leqslant \frac{1}{h}$$

at all points of $|z| < h$; but this inequality is satisfied at all other points of R_z as well, since at these points $|w| \leqslant 1$ and $|z| \geqslant h$. By interchanging the domains R_z and R_w we find that at all points w of R_w (i.e., since the correspondence is one-one at all points z of R_z) $|z/w| \leqslant 1/h$. Thus

$$h \leqslant \left| \frac{w(z)}{z} \right| \leqslant \frac{1}{h} \qquad \text{......(83·3)}$$

at all points of R_z.

But, if $|z| < hr$,

$$|w - z| = \left| \frac{w}{z} - 1 \right| |z| \leqslant hr \left| \frac{w}{z} - 1 \right|. \qquad \text{......(83·4)}$$

To obtain an upper bound for this expression we now write

$$z = ht, \quad \frac{w(z)}{z} = F(t), \quad \frac{F'(t)}{F(0)} = \phi(t), \qquad \text{......(83·5)}$$

so that, by (83·3),

$$h \leqslant |F(0)| \leqslant \frac{1}{h}, \quad h^2 \leqslant |\phi(t)| \leqslant \frac{1}{h^2}, \quad |F(0) - 1| \leqslant \frac{1 - h}{h},$$

$$\text{......(83·6)}$$

the last inequality holding since $F(0) = w'(0)$ was supposed real and positive. With this notation

$$\left|\frac{w}{z} - 1\right| = |F(t) - 1| \leqslant |F(t) - F(0)| + |F(0) - 1|,$$

and hence, by (83·6),

$$\left|\frac{w}{z} - 1\right| \leqslant |\phi(t) - 1| \frac{1}{h} + \frac{1-h}{h}. \qquad \ldots\ldots(83\cdot7)$$

84. If a is an arbitrary complex number, represented (say) by the point P, the number $|a - 1|$, being equal to the length of the segment UP, is not greater than the length UMP, where MP is an arc of a circle with centre O. We therefore have the inequality

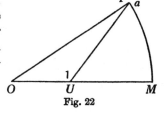

Fig. 22

$$|a - 1| \leqslant ||a| - 1| + |a||\text{I} \log a|. \qquad \ldots\ldots(84\cdot1)$$

85. By (84·1),

$$|\phi(t) - 1| \leqslant ||\phi(t)| - 1| + |\phi(t)||\text{I} \log \phi(t)|.$$

$$\ldots\ldots(85\cdot1)$$

But since, by (83·5), $\phi(0) = 1$, the branch of the function $\psi(t) = \log \phi(t)$, which vanishes for $t = 0$, satisfies, in consequence of (83·6), the inequality

$$|\text{R}\psi(t)| \leqslant -2 \log h;$$

and therefore, by § 76, if $|t| \leqslant r$,

$$|\text{I} \log \phi(t)| \leqslant \frac{-4 \log h}{\pi} \log \frac{1+r}{1-r}. \qquad \ldots\ldots(85\cdot2)$$

Also, by (83·6),

$$|\phi(t)| \leqslant \frac{1}{h^2}, \quad ||\phi(t)| - 1| \leqslant \frac{1-h^2}{h^2}. \qquad \ldots\ldots(85\cdot3)$$

It now follows from (85·1), (85·2) and (85·3) that

$$|\phi(t) - 1| \leqslant \frac{1-h^2}{h^2} - \frac{4 \log h}{\pi h^2} \log \frac{1+r}{1-r},$$

and on using (83·4) and (83·7) we obtain finally

$$|w - z| \leqslant m(h, r),$$
$$m(h, r) = \frac{(1 - h^3)r}{h^2} - \frac{4r \log h}{\pi h^2} \log \frac{4r}{1-r}. \qquad \Biggr\} \qquad \ldots\ldots(85\cdot4)$$

From this expression it is seen that not only is $\lim_{h \to 1} m(h, r) = 0$ for every fixed r, as was to be proved, but also $\lim_{h \to 1} m(h, h) = 0$.

Remark. By longer calculations than the above it can be shown that the function $m(h, r)$ of (85·4) can be replaced by a much smaller function.

86. Problem.

The reader may now make use of the results of §§ 82–85 to establish the following theorem.

Let R_u and S_u be two simply-connected domains of the u-plane which contain the point $u = 0$ in their interiors and which are represented on the interior of the unit-circle $|z| < 1$ by the functions $u = f(z)$, $(f(0) = 0,$ $f'(0) > 0)$, and $u = g(z)$, $(g(0) = 0, g'(0) > 0)$, respectively. It is supposed that the functions $u = f(z)$ and $u = g(z)$ represent the circle $|z| < h < 1$ on domains which lie entirely inside S_u and R_u respectively, i.e. that all points of R_u or of S_u whose non-Euclidean distances from $u = 0$ are less than $\frac{1}{2} \log \dfrac{1 + h}{1 - h}$ are common to both the domains. Then it is to be proved that, if $|z| \leqslant hr < h$, there is a relation $|f(z) - g(z)| \leqslant \mu(h, r)$, where $\lim_{h \to 1} \mu(h, r) = 0$. What is the geometrical interpretation of this relation?

87. Extensions of Schwarz's Lemma.

Let $w = f(z)$ again denote a function which is regular and such that $|f(z)| < 1$ at all points of the circle $|z| < 1$. Together with any non-Euclidean circle $C(z)$ with non-Euclidean centre z and non-Euclidean radius $\rho(z)$ we consider the circle $\Gamma(z)$ in the w-plane whose non-Euclidean centre and radius are $w = f(z)$ and $\rho(z)$. If A_z denotes the set of all points belonging to an arbitrary set of such circles $C(z)$ (and their interiors), and A_w that covered by the set of corresponding circles $\Gamma(z)$ in the w-plane, it follows from Theorem 3 of § 70 that any interior point of A_z is transformed by the function $w = f(z)$ into an interior point of A_w.

If the set of circles $C(z)$ is enumerable we can also consider the set of points B_z consisting of all points which lie in all but a finite number of the circles $C(z)$, and compare it with the set of points B_w obtained in the same way from the corresponding circles $\Gamma(z)$ in the w-plane. It follows from the theorem quoted that $w = f(z)$ transforms any point of B_z into a point of B_w.

Various applications of the above considerations can be made, and those applications are particularly interesting which throw light on the behaviour of $f(z)$ in the neighbourhood of the frontier $|z| = 1$.

88. Suppose first that the centre z of the circle $C(z)$ describes that diameter of the circle $|z| < 1$ which lies along the real axis, and that all the circles $C(z)$ have the same non-Euclidean radius. In this case the domain A_z, consisting of all points interior to any of these circles $C(z)$, is the area bounded by two circular arcs, and is symmetrical with respect to the real axis (see § 48).

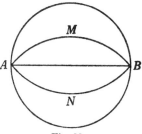

Fig. 23

If then $f(z)$ is a function which is regular and such that $|f(z)| < 1$ if $|z| < 1$, and if also $f(z)$ *is real whenever z is real*, the circles $\Gamma(z)$ in the w-plane corresponding to these circles $C(z)$ will have their non-Euclidean centres also on the real axis, and, since their non-Euclidean radii are the same as the non-Euclidean radii of the circles $C(z)$, must lie inside precisely the same area A_z, but in the w-plane. We therefore have the following theorem:

THEOREM. *Let $f(z)$ be a function which is regular and such that $|f(z)| < 1$ if $|z| < 1$; suppose also that $f(z)$ is real when z is real. Denoting by A and B the points -1 and $+1$, let AMB be a circular arc lying inside $|z| < 1$, and let ANB be the image of AMB in the real axis. Then if z has any value inside the area $AMBN$ the value of $f(z)$ also lies within this same area.*

89. Julia's Theorem.

The method of § 87 leads without difficulty to an important theorem which is due to *G. Julia*[16]. We note in the first place that two real points x, h $(x < h)$ in the circle $|z| < 1$ are at the same non-Euclidean distance as two real points y, k in the circle $|w| < 1$ if the cross-ratio of the numbers $(-1, x, h, 1)$ is equal to that of the numbers $(-1, y, k, 1)$, i.e. if

$$\frac{(1-x)(1+h)}{(1+x)(1-h)} = \frac{(1-y)(1+k)}{(1+y)(1-k)}; \qquad \ldots\ldots(89\cdot1)$$

or, writing $1 - h = u$, $1 - k = v$, if

$$\frac{1-y}{1+y} = \frac{v(2-u)}{u(2-v)}\frac{1-x}{1+x}. \qquad \ldots\ldots(89\cdot2)$$

Consider now two sequences of positive real numbers $u_1, u_2, u_3, \ldots$ and $v_1, v_2, v_3, \ldots$, satisfying the following conditions:

$$\lim_{n\to\infty} u_n = \lim_{n\to\infty} v_n = 0, \qquad \lim_{n\to\infty} \frac{v_n}{u_n} = \alpha, \qquad \ldots\ldots(89\cdot3)$$

where α is finite. Denote by K_n' the circle having the point $h_n = 1 - u_n$ as non-Euclidean centre and the point x on its circumference, and by Γ_n' a circle of the same non-Euclidean radius and having the point $k_n = 1 - v_n$ as non-Euclidean centre.

The circle Γ_n' cuts the real axis of the w-plane at a point y_n which is obtained by replacing u, v, y in (89·2) by u_n, v_n, y_n. The circles K_n'

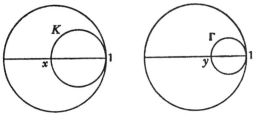

Fig. 24

converge, in consequence of (89·3), to an oricycle K of the non-Euclidean plane $|z| < 1$, passing through the points $z = 1$ and $z = x$, and the circles Γ_n' to an oricycle Γ of the non-Euclidean plane $|w| < 1$, passing through $w = 1$ and $w = y$; here y, being the limit of y_n, is given by the equation [see (89·2) and (89·3)]

$$\frac{1-y}{1+y} = \alpha \frac{1-x}{1+x} \qquad \ldots\ldots(89\cdot4)$$

If r and ρ denote the *Euclidean* radii of K and Γ, then $x = 1 - 2r$, $y = 1 - 2\rho$, so that, by (89·4),

$$\rho = \frac{\alpha r}{1 - r(1-\alpha)}. \qquad \ldots\ldots(89\cdot5)$$

90. Now let $f(z)$ be an analytic function which is regular and of modulus less than unity in the circle $|z| < 1$, and suppose further that there is a sequence of numbers $z_1, z_2, z_3, \ldots$ such that

$$\lim_{n\to\infty} z_n = 1, \quad \lim_{n\to\infty} f(z_n) = 1, \qquad \ldots\ldots(90\cdot1)$$

and

$$\lim_{n\to\infty} \frac{1 - |f(z_n)|}{1 - |z_n|} = \alpha, \qquad \ldots\ldots(90\cdot2)$$

where α is finite. It follows from (72·3) that

$$\alpha \geqslant \frac{1 - |f(0)|}{1 + |f(0)|} > 0. \qquad \ldots\ldots(90\cdot3)$$

Write now $u_n = 1 - |z_n|$, $v_n = 1 - |f(z_n)|$, and construct the circles K_n', Γ_n' of §89. Still using K and Γ to denote the oricycles to which K_n' and Γ_n' converge, let K_n and Γ_n denote circles with the same non-Euclidean radii as K_n' and Γ_n', but having their non-Euclidean centres at z_n and $f(z_n)$ respectively, instead of at $|z_n|$ and $|f(z_n)|$. Thus K_n is obtained from K_n' by a rotation about the origin; but since, by (90·1), the angle of rotation tends to zero as n tends to infinity, the circles K_n tend to the same limiting figure K as the circles K_n'. Similarly the circles Γ_n tend to Γ.

Thus, by § 87, if the point z lies inside K, the point $w = f(z)$ cannot lie outside Γ; we can in fact show, by an argument similar to that of § 67, that $f(z)$ must lie *inside* Γ.

The above result constitutes Julia's Theorem* (16).

91. The work of § 90 may be completed by adding that if there is a point z_1 of the frontier of K which is transformed into a point $f(z_1)$ of the frontier of Γ, then $f(z)$ must be a bilinear function. For, by § 48, any two oricycles which touch the circle $|z| = 1$ at $z = 1$ are parallel curves in the sense of non-Euclidean geometry; also equation (89·4), which represents a non-Euclidean motion, transforms the two oricycles through the points x_1 and x_2 $(x_1 < x_2)$ into the two oricycles through y_1 and y_2, and the distances between these two pairs of oricycles are the same, say, δ. If then z_1 is a point on the oricycle through x_1 such that the corresponding point $w_1 = f(z_1)$ lies on the oricycle through y_1, and we determine on the oricycle through x_2 the unique point z_2 for which $D(z_1, z_2) = \delta$, then, by Julia's Theorem, the point $w_2 = f(z_2)$ cannot lie outside the oricycle through y_2; that is to say, $D(w_1, w_2) \geqslant \delta$. But, by Theorem 3 of § 70, $D(w_1, w_2) \leqslant \delta$. Hence $D(w_1, w_2) = \delta$, and this cannot be true for a single pair of points unless $w = f(z)$ represents a non-Euclidean motion. Since also $f(1) = 1$, $f(z)$ must be of the form

$$f(z) = \frac{z - z_0}{1 - \bar{z}_0 z} \frac{1 - \bar{z}_0}{1 - z_0}. \qquad \ldots\ldots(91\cdot1)$$

92. Julia's Theorem may be interpreted geometrically as follows: in Fig. 25 we see that

$$\frac{AP}{PB} = \frac{Ax}{xE} = \frac{1 - x}{1 + x},$$

and similarly that

$$\frac{A_1 P_1}{P_1 B_1} \leqslant \frac{1 - y}{1 + y}.$$

Thus, if P, P_1 denote the points $z, f(z)$ respectively, Julia's Theorem gives, on using (89·4), the relation

$$\frac{A_1 P_1}{P_1 B_1} \leqslant \alpha \frac{AP}{PB}. \qquad \ldots\ldots(92\cdot1)$$

---

* In his own proof of this theorem Julia requires that $f(z)$ should be regular at $z = 1$. It is remarkable that it should be possible to prove the theorem without assuming regularity at $z = 1$.

But by elementary geometry

$$\frac{AP}{PB} = \frac{AP^2}{AP \cdot PB} = \frac{AP^2}{CP \cdot PD} = \frac{|1-z|^2}{1-|z|^2}, \quad \ldots\ldots(92 \cdot 2)$$

so that Julia's Theorem is equivalent to the following inequality:

$$\frac{|1-f(z)|^2}{1-|f(z)|^2} \leqslant \alpha \frac{|1-z|^2}{1-|z|^2}. \quad \ldots\ldots(92 \cdot 3)$$

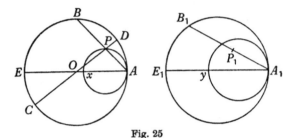

Fig. 25

93. A number of interesting consequences follow from Julia's Theorem. If we continue to denote, as in (89·5), the Euclidean radii of the circles K and Γ by r and ρ, then, by the theorem, the *real* point

$$x = 1 - 2r \quad \ldots\ldots(93 \cdot 1)$$

corresponds to a point $f(x)$ lying inside or on the frontier of Γ, i.e. such that

$$1 - |f(x)| \leqslant |1-f(x)| \leqslant 2\rho.$$

Thus, by (89·5) and (93·1),

$$\frac{1-|f(x)|}{1-x} \leqslant \frac{|1-f(x)|}{1-x} \leqslant \frac{\alpha}{1-r(1-\alpha)};$$

hence immediately

$$\varliminf_{x \to 1} \frac{1-|f(x)|}{1-x} \leqslant \varliminf_{x \to 1} \frac{|1-f(x)|}{1-x} \leqslant \alpha; \quad \ldots\ldots(93 \cdot 2)$$

$$\varlimsup_{x \to 1} \frac{1-|f(x)|}{1-x} \leqslant \varlimsup_{x \to 1} \frac{|1-f(x)|}{1-x} \leqslant \alpha. \quad \ldots\ldots(93 \cdot 3)$$

But α may be any number for which an equation such as (90·2) holds; thus the most favourable choice of α will be made by choosing the sequence z_n of (90·1) and (90·2) so that

$$\alpha = \lim_{x \to 1} \frac{1-|f(x)|}{1-x}.$$

When this is done, inequalities (93·2) and (93·3) at once establish the existence of the limits

$$\lim_{x \to 1} \frac{1 - |f(x)|}{1 - x} = \lim_{x \to 1} \frac{|1 - f(x)|}{1 - x} = \alpha. \qquad \ldots \ldots (93·4)$$

Since, by (90·3), $\alpha > 0$, the last equation shows that

$$\lim_{x \to 1} \frac{1 - |f(x)|}{|1 - f(x)|} = 1. \qquad \ldots \ldots (93·5)$$

But, if we write

$$1 - f(x) = \lambda e^{i\theta}, \qquad \ldots \ldots (93·6)$$

where $\lambda > 0$ and θ is real, we have

$$\frac{1 - |f(x)|}{|1 - f(x)|} = \frac{2 \cos \theta - \lambda}{1 + \sqrt{(1 - 2\lambda \cos \theta + \lambda^2)}}; \qquad \ldots \ldots (93·7)$$

also, when x tends to unity, λ tends to zero, and therefore, by (93·5) and (93·7), $\cos \theta$ (and hence also $e^{i\theta}$) must tend to unity. From this it follows, on using (93·5) and (93·6), that

$$\lim_{x \to 1} \frac{1 - f(x)}{1 - x} = \lim_{x \to 1} \frac{|1 - f(x)| e^{i\theta}}{1 - x} = \alpha. \qquad \ldots \ldots (93·8)$$

We may collect what has been proved in the form of the following theorem:

THEOREM. *If $f(z)$ is regular and such that $|f(z)| < 1$ in the circle $|z| < 1$, and if there exists inside the unit-circle a sequence of points z_1, z_2, ... for which $\lim z_n = 1$, $\lim f(z_n) = 1$, and*

$$\lim_{n \to \infty} \frac{1 - |f(z_n)|}{1 - |z_n|} \qquad \ldots \ldots (93·9)$$

is finite, then the following limit-equations hold, wherein x tends to its limit through arbitrary real values and α is a real positive constant

$$\lim_{x \to 1} \frac{1 - |f(x)|}{1 - x} = \lim_{x \to 1} \frac{|1 - f(x)|}{1 - x} = \lim_{x \to 1} \frac{1 - f(x)}{1 - x} = \alpha. \qquad \ldots \ldots (93·10)$$

CHAPTER V

THE FUNDAMENTAL THEOREMS OF CONFORMAL REPRESENTATION

94. Continuous convergence.

If A is an arbitrary set of points in the complex plane, we consider a sequence of complex functions $f_1(z)$, $f_2(z)$, ... which are defined and finite at all points of A; and we introduce the following *definition*: if z_0 is a limiting-point of A the sequence $f_n(z)$ will be said to be *continuously convergent* at z_0 if the limit

$$\lim_{n \to \infty} f_n(z_n) = f(z_0) \qquad \ldots\ldots(94 \cdot 1)$$

exists for every sequence of points z_n of A having z_0 as its limit.

It is readily proved that the limit $f(z_0)$ is independent of the particular sequence z_n, and that every sub-sequence f_{n_1}, f_{n_2}, ... of the given sequence also converges *continuously* to $f(z_0)$ at the point z_0, i.e. that

$$\lim_{k \to \infty} f_{n_k}(z_k) = f(z_0), \quad (n_1 < n_2 < \ldots). \qquad \ldots\ldots(94 \cdot 2)$$

The following theorem also follows immediately from the definition of continuous convergence:

THEOREM. *If* (i) *the sequence of functions* $w = f_n(z)$ *is defined in the set of points A and converges continuously at the point z_0 to the finite limit* $w_0 = f(z_0)$, (ii) *for all points z of A and for all values of n the point* $w = f_n(z)$ *belongs to a set of points B of the w-plane, and* (iii) *the sequence of functions $\phi_n(w)$ is defined in B and converges continuously at w_0, then the sequence*

$$F_n(z) = \phi_n\{f_n(z)\}$$

converges continuously at z_0.

95. Limiting oscillation.

Let $f_1(z), f_2(z)$, ... be an arbitrary sequence of complex functions which are defined in a domain R and uniformly bounded in a neighbourhood of every point z_0 of R. Denoting by z_0 a point of R, let $C^{(1)}$, $C^{(2)}$, ... be a sequence of circles lying in R, having their centres at z_0, and such that their radii decrease steadily to the limit zero. Further, let $O_n^{(k)}$ denote the oscillation of $f_n(z)$ in the circle $C^{(k)}$, i.e.,

$$O_n^{(k)} = \sup |f_n(z') - f_n(z'')|, \qquad \ldots\ldots(95 \cdot 1)$$

where z', z'' are any two points of the domain $C^{(k)}$. If we now write

$$\omega^{(k)} = \overline{\lim_{n \to \infty}} \ O_n^{(k)}, \qquad \ldots\ldots(95\text{·}2)$$

we have, since $C^{(k+1)} \subset C^{(k)}$, $\omega^{(k+1)} \leqslant \omega^{(k)}$; hence the limit

$$\omega(z_0) = \lim_{k \to \infty} \omega^{(k)}$$

exists. We see also that this limit depends only on the sequence $f_n(z)$ and on the point z_0, *but not on the choice of the circles* $C^{(k)}$.

The number $\omega(z_0)$ is called the *limiting oscillation* of the sequence $f_n(z)$ at the point z_0.

96. We now prove the following theorems:

THEOREM 1. *If $\omega(z_0)$ and $\sigma(z_0)$ denote respectively the limiting oscillations of a sequence of functions $f_n(z)$ and of a sub-sequence $f_{n_r}(z)$ at the point z_0, then*

$$\sigma(z_0) \leqslant \omega(z_0). \qquad \ldots\ldots(96\text{·}1)$$

For, by (95·2), for every circle $C^{(k)}$

$$\sigma^{(k)} \leqslant \omega^{(k)}.$$

THEOREM 2. *If at a point z_0 the limiting oscillation $\omega(z_0) > 0$, the sequence $f_n(z)$ cannot converge continuously at z_0.*

For, by hypothesis, for every integer k an infinite number of functions of the sequence have, in the circle $C^{(k)}$, an oscillation greater than $\omega^{(k)} - \tfrac{1}{2}\omega(z_0)$, i.e. (since $\omega^{(k)} \geqslant \omega(z_0)$) greater than $\tfrac{1}{2}\omega(z_0)$. Hence there are two sequences of points z_1', $z_2' \ldots$, and z_1'', $z_2'' , \ldots$, and also an increasing sequence of integers $n_1 < n_2 < \ldots$ such that simultaneously

$$\lim_{k \to \infty} z_k' = z_0, \quad \lim_{k \to \infty} z_k'' = z_0, \quad |f_{n_k}(z_k') - f_{n_k}(z_k'')| > \tfrac{1}{2}\omega(z_0).$$
$$\ldots\ldots(96\text{·}2)$$

Hence, of the two sequences

$$w_k' = f_{n_k}(z_k'), \quad w_k'' = f_{n_k}(z_k''), \qquad \ldots\ldots(96\text{·}3)$$

either at least one is not convergent, or, if both converge, their limits are unequal; and each of these possibilities excludes continuous convergence, by § 94.

THEOREM 3. *The sequence $f_n(z)$ converges continuously at z_0 provided that $\omega(z_0) = 0$ and that every neighbourhood of z_0 contains a point ζ such that the sequence $f_n(\zeta)$ is convergent.*

Thus, in particular, the sequence converges continuously if $\omega(z_0) = 0$ and $\lim_{n \to \infty} f_n(z_0)$ exists and is finite.

Let $z_1, z_2, \ldots$ denote an arbitrary sequence of points of R having z_0 as its limit. We have to show that, when the hypotheses of Theorem 3 are satisfied, the sequence

$$w_n = f_n(z_n) \qquad \ldots\ldots(96\cdot4)$$

is convergent. Denoting by ϵ an arbitrary positive number, we can, since $\lim_{k\to\infty} \omega^{(k)} = 0$, define a circle $C^{(k)}$ for which

$$\omega^{(k)} < \tfrac{1}{6}\epsilon. \qquad \ldots\ldots(96\cdot5)$$

In $C^{(k)}$ there is, by hypothesis, a point ζ at which the sequence $f_n(z)$ converges; and hence there exists a number N_1 such that, if $n > N_1$ and $m > N_1$,

$$|f_n(\zeta) - f_m(\zeta)| < \tfrac{1}{3}\epsilon. \qquad \ldots\ldots(96\cdot6)$$

Further, by (95·2) and (96·5) there is a number N_2 such that if $n > N_2$ the oscillation $O_n^{(k)} < \tfrac{1}{3}\epsilon$; finally, suppose that for $n > N_3$ the point z_n lies in $C^{(k)}$. Thus if N denotes the greatest of the numbers N_1, N_2, N_3, we have, provided only that $n > N$ and $m > N$, in addition to (96·6), the inequalities

$$|f_n(z_n) - f_n(\zeta)| < \tfrac{1}{3}\epsilon, \quad |f_m(z_m) - f_m(\zeta)| < \tfrac{1}{3}\epsilon. \quad \ldots\ldots(96\cdot7)$$

In view of (96·4) to (96·6),

$$|w_n - w_m| < \epsilon \quad (n, m > N). \qquad \ldots\ldots(96\cdot8)$$

Hence, by Cauchy's criterion, the sequence w_n converges, so that Theorem 3 is proved.

THEOREM 4. *If B denotes the set of points z of A at which the sequence $f_n(z)$ converges to a function $f(z)$, then every limiting point z_0 of B such that $\omega(z_0) = 0$ belongs to B, and the function $f(z)$, which is defined in B, is continuous at z_0.*

From Theorem 3 we have at once not only that the functions $f_n(z)$ converge at z_0, so that z_0 is a point of B, but also that the convergence at z_0 is continuous. Now consider an arbitrary sequence of points $z_1, z_2, \ldots$ of B, having z_0 as limit; to establish the continuity of $f(z)$ at z_0 we have only to prove that

$$\lim_{n\to\infty} f(z_n) = f(z_0). \qquad \ldots\ldots(96\cdot9)$$

Since the sequence $f_n(z)$ converges at each of the points z_k, we can find an increasing sequence of integers $n_1, n_2, \ldots$, such that for every k

$$|f_{n_k}(z_k) - f(z_k)| < \frac{1}{k}. \qquad \ldots\ldots(96\cdot10)$$

Equation (96·9) now follows on combining (96·10) with the fact that, on account of the continuous convergence of the sequence at z_0,

$$\lim_{k\to\infty} f_{n_k}(z_k) = f(z_0). \qquad \ldots\ldots(96\cdot11)$$

97. Normal families of bounded functions (17) (18).

We consider a definite class $\{f(z)\}$ of functions $f(z)$, which are defined in a domain R and are uniformly bounded in certain neighbourhoods of every point z_0 of R. For such a family of functions we can define for each point z_0 of R a number $\omega(z_0)$, which we shall call the *limiting oscillation* of the family. We again consider the sequence of circles $C^{(k)}$ defined in § 95, and we denote by $\omega^{(k)}$ the upper limit of the oscillations in $C^{(k)}$ of all functions $f(z)$ of the given family; thus $\omega^{(k)}$ has the following two properties: if $p > \omega^{(k)}$ at most a *finite* number of distinct functions belonging to $\{f(z)\}$ have oscillations in $C^{(k)}$ which are greater than p, whereas if $q < \omega^{(k)}$ an infinite number of distinct functions of the family have in $C^{(k)}$ oscillations greater than q. Finally, we define the limiting oscillation by the equation

$$\omega(z_0) = \lim_{k \to \infty} \omega^{(k)}. \qquad \qquad \text{......(97·1)}$$

The following theorem is obvious:

THEOREM 1. *The limiting oscillation at z_0 of any sequence of functions $f_1(z), f_2(z), \ldots$ which belong to $\{f(z)\}$ cannot be greater than the limiting oscillation $\omega(z_0)$ of the family $\{f(z)\}$ at z_0.*

98. We now introduce the following *definition*: *a family $\{f(z)\}$ of functions defined in R and uniformly bounded in detail* (im kleinen) *is said to be normal on R if the limiting oscillation $\omega(z_0)$ vanishes at every point z_0 of R.*

We have

THEOREM 2. *If $f_1(z), f_2(z), \ldots$ is a sequence of functions all belonging to a family $\{f(z)\}$ which is normal on R, then the given sequence contains a sub-sequence $f_{n_1}(z), f_{n_2}(z), \ldots$ which is continuously convergent at every point of R* (19).

Consider a countable sequence of points $z_1, z_2, \ldots$ of R which lie everywhere dense on R. By the diagonal process (*Cantor* and *Hilbert*), a sub-sequence $f_{n_1}(z), f_{n_2}(z), \ldots$ can be picked out from the given sequence in such a way that, at each point of the countably infinite set, the limit

$$\lim_{k \to \infty} f_{n_k}(z_p) \qquad \qquad \text{......(98·1)}$$

exists. The conditions of Theorem 3, § 96, are satisfied by this sub-sequence at every point of R; this proves the present theorem.

99. It is easy to show that the two conceptions, (*a*) *continuous convergence* of a sequence of *continuous* functions at *all* points of a *closed* set γ, and (*b*) *uniform convergence* of this sequence on γ, are completely equivalent (18).

The following theorem is a consequence of this equivalence, but we shall give a direct proof.

THEOREM 3. *If a sequence* $f_1(z)$, $f_2(z)$, ... *of functions converges continuously to the function* $f(z)$ *at all points of a bounded closed set* γ, *and if* $f(z) \neq 0$ *on* γ, *then there is a positive number* m *and a positive integer* N, *such that, at every point of* γ *and for each* $n > N$,

$$|f_n(z)| > m. \qquad \ldots\ldots(99\cdot1)$$

If this were not so, it would be possible to find an increasing sequence of integers, $n_1 < n_2 < \ldots$, and a sequence of points z_1, z_2, ... of γ, such that

$$|f_{n_k}(z_k)| < \frac{1}{k} \quad (k = 1, 2, \ldots). \qquad \ldots\ldots(99\cdot2)$$

By omitting some of the points z_k and ordering afresh those that remain we can ensure that

$$\lim_{k \to \infty} z_k = z_0. \qquad \ldots\ldots(99\cdot3)$$

Since γ is closed, z_0 belongs to γ and therefore the sequence $f_1(z)$, $f_2(z)$, ... converges continuously at z_0. Hence

$$f(z_0) = \lim_{k \to \infty} f_{n_k}(z_k) = 0, \qquad \ldots\ldots(99\cdot4)$$

and a hypothesis is contradicted. This proves the theorem.

100. Existence of the solution in certain problems of the calculus of variations.

Weierstrass' Theorem that a continuous function defined on a bounded closed set in n-dimensional space attains its maximum at some point of the set can be extended, in certain circumstances, to function-space, that is, to space in which any element is defined not by a system of n numbers but by a definite function. For instance, suppose that a family $\{f(z)\}$ of complex functions is given, all the functions being defined in a domain R. The family is said to be *compact* if every sequence $f_1(z)$, $f_2(z)$ ..., of functions of the family, contains a sub-sequence $f_{n_1}(z), f_{n_2}(z)$, ... which converges in R to a function $f_0(z)$, where $f_0(z)$ also belongs to the family (20).

Secondly, we consider a functional $J(f)$, by means of which a finite number is associated with each member $f(z)$ of $\{f(z)\}$. The functional $J(f)$ is said to be *continuous* if the convergence in R of the sequence $f_1(z), f_2(z), \ldots$ to the function $f_0(z)$ of $\{f(z)\}$ always implies $J(f_n) \rightarrow J(f_0)$. We now have:

THEOREM. *If $J(f)$ is a continuous functional defined in the compact family $\{f(z)\}$, then the problem*

$$| J(f) | = \text{Maximum} \qquad \ldots\ldots(100\cdot1)$$

has a solution within the family $\{f(z)\}$.

That is to say, there is at least one member $f_0(z)$ of the family $\{f(z)\}$ such that *all* members $f(z)$ of $\{f(z)\}$ satisfy

$$| J(f) | \leqslant | J(f_0) |. \qquad \ldots\ldots(100\cdot2)$$

For if α denotes the finite (or infinite) upper bound of $| J(f) |$, as $f(z)$ varies in the family $\{f(z)\}$, then there are sequences $f_1(z), f_2(z), \ldots$ of members of $\{f(z)\}$, such that

$$\lim_{n \to \infty} | J(f_n) | = \alpha. \qquad \ldots\ldots(100\cdot3)$$

Now, since $\{f(z)\}$ is compact, a sub-sequence $f_{n_1}(z), f_{n_2}(z), \ldots$ can be selected from the sequence $f_1(z), f_2(z), \ldots$ in such a way that the sub-sequence converges to a member $f_0(z)$ of $\{f(z)\}$. The continuity of $J(f)$ shows that

$$| J(f_0) | = \lim_{k \to \infty} | J(f_{n_k}) | = \alpha.$$

This not only proves the theorem but also shows that α is finite.

101. Normal families of regular analytic functions.

We now consider families $\{f(z)\}$ whose members $f(z)$ are all defined, analytic and regular, in the interior of a domain R. For these families the following theorem holds:

THEOREM 1. *If $\{f(z)\}$ is a family of analytic functions, which are all regular in the interior of a domain R, and which are uniformly bounded on the circumference of any circle lying strictly within R, then $\{f(z)\}$ is normal in R, and the same is true of the family $\{f'(z)\}$ of derived functions, obtained from the functions $f(z)$.*

Let z_0 be a point of R and $|z - z_0| \leqslant \rho$ a closed circle lying strictly within R. There is a positive number M such that all members of $\{f(z)\}$ satisfy the relation $| f(z) | < M$ at all points of $|z - z_0| = \rho$ and hence also at all points of $|z - z_0| < \rho$.

Account being taken of Theorem 5, § 71, it is seen that, if r is any number between 0 and 1, the relation

$$|f(z') - f(z'')| \leqslant \frac{M |z' - z''|}{\rho (1 - r^2)} < \frac{2Mr}{1 - r^2} \qquad \ldots\ldots(101\cdot1)$$

is satisfied by every pair (z', z'') of points of the circle $|z - z_0| < \rho r$ and by every member of $\{f(z)\}$. From this it follows immediately that the limiting oscillation of $\{f(z)\}$ vanishes at the point z_0. By § 98 the family $\{f(z)\}$ is normal.

To prove that the family $\{f'(z)\}$, consisting of the derived functions of the given functions, is also normal, we need only remark that by Theorem 4, § 71, the derived functions $f'(z)$ are uniformly bounded in the circle $|z - z_0| \leqslant \rho r$.

102. We shall now show that the boundary function $f_0(z)$ of a convergent sequence $f_1(z), f_2(z), \ldots$ of functions of the family is analytic. For this, as is well known, it is sufficient to show that $f_0(z)$ is differentiable at every point of R.

Take a closed circle $|z - z_0| \leqslant \rho$, lying in R, and in it consider the sequence of regular analytic functions $\phi_n(z)$, which is defined by the equations

$$\phi_n(z) = \frac{f_n(z) - f_n(z_0)}{z - z_0}, \quad (z \neq z_0), \qquad \ldots\ldots(102\cdot1)$$

$$\phi_n(z_0) = f_n'(z_0). \qquad \ldots\ldots(102\cdot2)$$

The sequence of functions $\phi_n(z)$ is uniformly bounded on the circumference $|z - z_0| = \rho$ of the circle in question, and hence, by the result of the preceding paragraph, it is normal. The sequence obviously converges at all points of the circle other than z_0, and therefore it converges continuously at the point z_0 itself (Theorem 3, § 96). Let the boundary function be denoted by $\phi(z)$; then (102·1) and (102·2) yield

$$\phi(z) = \frac{f_0(z) - f_0(z_0)}{z - z_0}, \quad (z \neq z_0), \qquad \ldots\ldots(102\cdot3)$$

$$\phi(z_0) = \lim_{n \to \infty} f_n'(z_0). \qquad \ldots\ldots(102\cdot4)$$

Further, by Theorem 4, § 96, $\phi(z)$ is continuous at the point z_0. The above equations therefore assert that

$$f_0'(z_0) = \phi(z_0) = \lim_{n \to \infty} f_n'(z_0). \qquad \ldots\ldots(102\cdot5)$$

This gives

THEOREM 2. *If a sequence of analytic functions is uniformly bounded in a domain in which the sequence converges, then the boundary function is analytic and its derived function is the limit of the derived functions of the functions of the approximating sequence.*

Keeping this theorem in mind, we shall, for the sake of brevity, describe a convergent sequence of regular functions which is uniformly bounded in detail as *regularly convergent.*

Suppose that a given sequence of analytic functions, defined in a domain R, is uniformly bounded in detail; the sequence is regularly convergent in R provided only that the points at which it converges possess a limiting point in the interior of R. For, if the given sequence did not converge everywhere, it would be possible to select from it two sub-sequences converging to two distinct analytic functions $f(z)$ and $g(z)$. But, at each point at which the given sequence converges, $(f(z) - g(z)) = 0$. These zeros cannot possess a limiting point within R.

103. The following theorem relating to regularly convergent sequences of analytic functions is a special case of a well-known theorem due to *Hurwitz*:

THEOREM 3. *If, in a domain R, the sequence of functions $f_1(z), f_2(z), \ldots$ converges regularly to the function $f(z)$, none of the functions $f_n(z)$ vanishing at any point of R, then either $f(z) \equiv 0$ or $f(z)$ does not vanish at any point of R.*

For suppose that $f(z)$ does not vanish identically in R. Corresponding to each point z_0 of R there is at least one circle $|z - z_0| \leqslant \rho$ lying within R and such that on its circumference $f(z) \neq 0$.

By Theorem 3 of § 99, there must be a number $m > 0$ such that

$$|f_n(z)| > m \qquad \ldots\ldots(103\cdot1)$$

for all points on the circle $|z - z_0| = \rho$, provided that n is sufficiently large. But in this circle $f_n(z) \neq 0$, and therefore, if n is sufficiently large,

$$|f_n(z_0)| \geqslant m. \qquad \ldots\ldots(103\cdot2)$$

Thus $\qquad |f(z_0)| = \lim_{n \to \infty} |f_n(z_0)| \geqslant m,$

and it follows that $f(z_0) \neq 0$.

An immediate corollary of this theorem is:

THEOREM 4. *If in a domain R a sequence of functions $f_1(z), f_2(z), \ldots$ converges regularly to a function $f(z)$ which is not a constant, then any neighbourhood N_{z_0} of a point z_0 of R contains points z_n such that*

$$f_n(z_n) = f(z_0),$$

if n is sufficiently large.

If this were not so, the given sequence would yield an infinite sub-sequence, $f_{n_1}(z), f_{n_2}(z), \ldots$, such that, in the neighbourhood N_{z_0} of z_0, the functions $(f_{n_k}(z) - f(z_0))$ all differ from zero. Since the boundary function $(f(z)-f(z_0))$ of this last sequence vanishes at the point z_0, Theorem 3 shows that $(f(z)-f(z_0))$ vanishes identically, i.e. the function $f(z)$ is a constant, and a hypothesis is contradicted.

104. Application to conformal representation.

The following theorem is of fundamental importance in the theory of conformal representation:

THEOREM. *If $f_1(z)$, $f_2(z)$, ... is a sequence of functions which converges regularly in a domain R, and if the functions give conformal transformations of R into simple domains S_1, S_2, ... respectively, which are uniformly bounded, then, either the boundary function $f(z)$ is a constant, or it gives a conformal transformation of R into a simple domain S.*

By hypothesis, $f_n(z) \neq f_n(z_0)$ when z and z_0 are points of R, $z \neq z_0$. The functions $\phi_n(z) = f_n(z) - f_n(z_0)$ do not vanish in the pricked (*punktiert*) domain $R - z_0$. These functions converge continuously in this domain, towards the function $\phi(z) = f(z) - f(z_0)$. By Theorem 3 of § 103 either $\phi(z)$ is identically zero, and $f(z)$ is then a constant, or $\phi(z)$ is different from zero and so $f(z) \neq f(z_0)$.

105. The main theorem of conformal representation (21).

Let R be an arbitrary bounded domain in the z-plane, containing the point $z = 0$ and therefore also a circular area κ defined by

$$|z| < \rho \qquad \text{......(105·1)}$$

in its interior. No assumption is made as to the connectivity of R.

Consider a family $\{f(z)\}$ of functions which are regular in the circle (105·1). The family is assumed to be made up of the function $f(z) \equiv 0$ and also all functions $f(z)$ which satisfy the following conditions:

(a) $f(0) = 0$,

(b) analytic continuation of $f(z)$ is possible along every path γ within R and the function $f(z)$ is always regular,

(c) if γ' and γ'' are two paths joining $z = 0$ to the points z' and z'' respectively, and if $_{\gamma'}F(z')$ and $_{\gamma''}F(z'')$ are the values obtained at z' and z'' by continuing $f(z)$ along these paths, then if $z' \neq z''$

$$_{\gamma'}F(z') \neq _{\gamma''}F(z'') \qquad \text{.....(105·2)}$$

(in particular, $_{\gamma}F(z) \neq 0$ provided $z \neq 0$),

(d) with the above notation

$$|_{\gamma}F(z)| < 1. \qquad \text{......(105·3)}$$

106. It will first be proved that the family $\{f(z)\}$ is compact (§ 100).

Condition (d) shows that any sequence of functions of the family contains a sub-sequence $f_1(z)$, $f_2(z)$, ... which satisfies the relation

$$\lim_{n \to \infty} f_n(z) = f_0(z) \qquad \text{......(106·1)}$$

in κ. We have to show that $f_0(z)$ belongs to the family $\{f(z)\}$.

It is obvious that $f_0(0) = 0$, so that either condition (a) is fulfilled or $f(z) \equiv 0$.

To verify condition (b) we must show that if γ' is a path within R joining $z = 0$ to a point z', and if for every point ζ of γ', other than z', the analytic continuation of $f_0(z)$ gives a function $_{\gamma'}F_0(\zeta)$ which exists, is regular and can be obtained as the limit of the analytic continuations $_{\gamma'}F_n(\zeta)$ of the functions $f_n(z)$, then all these conditions are satisfied at the point z' itself and in a certain neighbourhood of that point.

It is easy to show this, for the functions $_{\gamma'}F_n(z)$ are regular in a certain neighbourhood of z' and by condition (d) they form a normal family which converges to $_{\gamma'}F_0(z)$ at all points of a certain portion of γ' (§ 102).

To prove condition (c), consider two paths γ' and γ'' with distinct endpoints z' and z'', and let $N_{z'}$ and $N_{z''}$ be non-overlapping neighbourhoods of z' and z'' respectively. By Theorem 4, § 103, there are points z_n' in $N_{z'}$ and z_n'' in $N_{z''}$ such that the equations

$$_{\gamma'}F_n(z_n') = {}_{\gamma'}F_0(z'), \quad _{\gamma''}F_n(z_n'') = {}_{\gamma''}F_0(z'') \qquad \text{......(106·2)}$$

hold simultaneously, n being suitably chosen. By hypothesis the terms on the left in these equations are unequal. The required result,

$$_{\gamma'}F_0(z') \neq {}_{\gamma''}F_0(z''),$$

follows.

Finally, it is obvious that $f_0(z)$ satisfies condition (d).

107. Consider now a particular function $f(z)$ of the family and its analytic continuations $_{\gamma}F(z)$ in R. Suppose that there is a number w_0, where

$$w_0 = he^{i\theta}, \quad (0 < h < 1, \ \theta \text{ real}), \qquad \text{......(107·1)}$$

such that, for all points z of R and all possible paths γ joining the origin to the point z,

$$_{\gamma}F(z) \neq w_0. \qquad \text{......(107·2)}$$

We construct, in succession, the functions

$$f_1(z) = e^{-i\theta} f(z), \qquad \text{......(107·3)}$$

$$\alpha(z) = \frac{h - f_1(z)}{1 - h f_1(z)}, \qquad \text{......(107·4)}$$

$$\beta(z) = \sqrt{\alpha(z)}, \quad (\beta(0) = + \sqrt{h}), \qquad \text{......(107·5)}$$

$$g(z) = \frac{\sqrt{h} - \beta(z)}{1 - \sqrt{h}\,\beta(z)}, \qquad \text{......(107·6)}$$

and proceed to investigate their properties.

The function $f_1(z)$ clearly belongs to the family $\{f(z)\}$, and neither the function nor its analytic continuations takes the value h. The function $\alpha(z)$ is connected with $f_1(z)$ by a Möbius' transformation which transforms the unit-circle into itself. The conditions (b), (c), (d) of § 105 are all satisfied by $\alpha(z)$. The same is true of $\beta(z)$. Neither $\alpha(z)$ nor its continuations take the value zero, so that $\beta(z)$ is regular on every path γ. To verify condition (c) we notice that if analytic continuation of $\beta(z)$ along paths γ' and γ'' leads to coincident values at z' and z'', the same must hold for $\alpha(z)$, which would lead to the conclusion that $z' = z''$.

Finally we see, from $(107\cdot6)$, that $g(z)$ satisfies the conditions (b), (c), and (d), and further, since $\beta(0) = \sqrt{h}$, the function $g(z)$ vanishes at $z = 0$, so that it is a member of the family $\{f(z)\}$.

108. If the equations $(107\cdot3)$ to $(107\cdot6)$ are solved, $\beta(z)$, $\alpha(z)$, $f_1(z)$ being successively determined as functions of

$$g(z) = t, \qquad \dots\dots(108\cdot1)$$

it is found that

$$f(z) = e^{i\theta} t \frac{2\sqrt{h} - (1+h)t}{(1+h) - 2\sqrt{h}\,t}. \qquad \dots\dots(108\cdot2)$$

The expression on the right-hand side of this equation coincides with one of the functions investigated in § 56. It satisfies all the conditions of Schwarz's Lemma (as can also be verified by direct calculation) and therefore, for every value of t in the domain $0 < |t| < 1$,

$$\left| t \frac{2\sqrt{h} - (1+h)t}{(1+h) - 2\sqrt{h}\,t} \right| < |t|. \qquad \dots\dots(108\cdot3)$$

From $(108\cdot1)$ to $(108\cdot3)$ it follows that

$$|f(z)| < |g(z)| \qquad \dots\dots(108\cdot4)$$

at all points of the circle κ except the centre.

The results so far obtained may be summarised as follows:

THEOREM 1. *The family $\{f(z)\}$ is compact. If the family contains a function $f(z)$ which is not a constant but which is such that neither it nor its analytic continuations in R take the value w_0, where $|w_0| < 1$, then there is a member $g(z)$ of the family such that $(108\cdot4)$ holds whenever $0 < |z| < \rho$.*

109. Let z_1 be a fixed point of κ, $z_1 \neq 0$, and define a continuous functional $J(f)$ (see § 100), by means of the equation

$$J(f) = |f(z_1)|. \qquad \dots\dots(109\cdot1)$$

Since the family $\{f(z)\}$ is compact, § 100 shows that it has at least one member $f_0(z)$ such that

$$|f_0(z_1)| \geqslant |f(z_1)| \qquad \ldots\ldots(109\cdot2)$$

as $f(z)$ varies in the family. The domain R being bounded, the function $f(z) = \lambda z$ belongs to the family, if λ is a positive number taken sufficiently small, and thus, from $(109\cdot2)$,

$$|f_0(z_1)| \geqslant \lambda |z_1| > 0,$$

which shows that $f_0(z)$ is not a constant.

Using the theorem of § 108 we obtain at once:

THEOREM 2. *The family* $\{f(z)\}$ *contains at least one function* $f_0(z)$ *which, with its analytic continuations, takes every value in the circular area* $|w| < 1$.

110. We now consider functions $f(z)$ which satisfy conditions (b), (c) and (d) but not necessarily condition (a) of § 105, and which have in addition the following property:

(e) To every point w_0 of the unit-circle $|w| < 1$ there corresponds a path γ_0, joining $z = 0$ and some point z_0, such that, with the notation used above,

$$_{\gamma_0}F(z_0) = w_0. \qquad \ldots\ldots(110\cdot1)$$

The last theorem proves the existence of such functions. Further, by the property (c), the number z_0 in $(110\cdot1)$ is a single-valued function of w_0, defined throughout the interior, $|w| < 1$, of the unit-circle. This function is the inverse function of $_{\gamma_0}F(z) = w$ in some neighbourhood of each point w_0, and it is therefore regular. Thus it is seen that the function, which we denote by $\phi(w)$, is single-valued, regular and analytic throughout the circle $|w| < 1$ (§ 63). The equation

$$z = \phi(w) \qquad \ldots\ldots(110\cdot2)$$

therefore gives a conformal transformation of the circular area $|w| < 1$ into a Riemann surface which covers the whole domain R.

111. A second function $g(z)$, satisfying (b), (c), (d) and (e), yields an inverse function

$$z = \psi(\omega) \qquad \ldots\ldots(111\cdot1)$$

which has the same properties as the function $\phi(w)$ of $(110\cdot2)$. By means of $(110\cdot2)$ a correspondence is set up between the points w_0 within the unit-circle and points z_0, with associated paths γ_0. The paths may be obtained, for instance, as the paths in the z-plane resulting from the transformation of straight lines joining 0 to w_0. If we write

$$\omega_0 = {}_{\gamma_0}G(z_0),$$

then ω_0 is a single-valued regular function of w_0. In what has just been said ω and w are interchangeable, and it follows from § 68 that the equation $\phi(w) = \psi(\omega)$ gives a non-Euclidean motion of the unit-circle on itself.

112. The monodromy theorem (§ 63) shows that if the given domain R is simply-connected the inverse of the function (110·2) must be a regular *single-valued* function of z in R. The equation $z = \phi(w)$ then gives a one-one conformal transformation of the interior of the unit-circle on the domain R.

It has already been shown (§ 64) that any simply-connected domain with more than one frontier-point can be transformed conformally into a bounded domain. Combining this with the results just obtained, we have

THEOREM 3. *The interior of any simply-connected domain R with more than one frontier-point can be represented on the interior of the unit-circle by means of a one-one conformal transformation. The representation is not unique. A given directed line-element in R can be made to correspond to an arbitrary line-element in* $|w| < 1$*. When such a correspondence is assigned the transformation is determined uniquely.*

113. On the other hand, if the inverse of the function (110·2) is a single-valued function of z in R, then by § 63 the domain R is simply-connected. Thus if R is *multiply*-connected there must be at least two points w' and w'' in $|w| < 1$ such that $\phi(w') = \phi(w'')$. Hence there is at least one non-Euclidean motion of the circle $|w| < 1$ for which $\phi(w)$ is invariant (§ 111). The aggregate of such non-Euclidean motions forms a group, and $\phi(w)$ is an *automorphic function*.

The points w of the circle $|w| < 1$ such that $\phi(w) = \phi(w_0)$ are called the points equivalent to w_0. If w_0 is any point of $|w| < 1$ there is a neighbourhood N_{w_0} of w_0 for which $z = \phi(w)$ gives a one-one transformation into a portion of R. No two points of N_{w_0} can be equivalent, and it follows that no member of the group of non-Euclidean motions is a non-Euclidean rotation (§ 45) about w_0 as fixed point. Since w_0 was an arbitrary point of $|w| < 1$, the group contains only limit-rotations and non-Euclidean translations (§ 46). Repetition of a non-Euclidean translation never leads back to the original figure. The same is true of a limit-rotation, and the group is therefore *infinite*. The points equivalent to a given point of the circular area $|w| < 1$ form an infinite set.

THEOREM 4. *If the domain R is multiply-connected, then the function* (110·2) *is automorphic, and it is invariant for certain transformations whose aggregate forms an infinite group of non-Euclidean translations and limit-rotations.*

114. In the case where R is doubly-connected it is easy to form the group of transformations.

Suppose first that R is a simply-connected domain from which the single point z_0 has been removed. Then a one-one conformal transformation can be found, to transform R into the pricked (*punktiert*) circular domain

$$0 < |t| < 1. \qquad \qquad \ldots\ldots(114\cdot1)$$

By means of elementary functions it can then be shown that the group is a cyclic group of limit-rotations.

The following very useful observation, first made by T. Radó, is of service in this connection :

Let $z = \phi(t)$ be a single-valued, regular, analytic function defined in (114·1) and bounded in some neighbourhood of $t = 0$. Suppose also that, if t_1 and t_2 are any two distinct points of (114·1), then $\phi(t_1) \neq \phi(t_2)$. By Riemann's Theorem, there is a number z_0 such that on writing $\phi(0) = z_0$ the function $\phi(t)$ obtained is regular and analytic in the whole circular area $|t| < 1$. Also, for $t \neq 0$, $\phi(t) \neq z_0$. For if $\phi(t') = z_0$, then $z = \phi(t)$ transforms a neighbourhood of t' into a neighbourhood of z_0. From this it is readily seen that it would then be possible to satisfy the equation $\phi(t_1) = \phi(t_2)$, where t_1 and t_2 are two distinct points of (114·1). Thus the function $z = \phi(t)$ gives a one-one conformal transformation of (114·1) into a pricked simply-connected domain.

Now assume that the points equivalent to a point w_0, where $|w_0| < 1$, are generated by a cyclic group of limit-rotations. A Möbius transformation can be found which transforms the circle $|w| < 1$ into the half-plane $\Re u < 0$, while at the same time the limit-rotation from which the group is generated becomes a Euclidean translation determined by the vector $2\pi i$.

If now $t = e^u$, the domain R is represented on the pricked circular area $0 < |t| < 1$ by a one-one conformal transformation. Then, by Radó's result, R coincides with a simply-connected domain from which a single point has been removed.

115. Next let the frontier of R consist of two distinct continua C_1 and C_2, each containing more than one point. Regard C_2 as the frontier of a simply-connected domain containing C_1. This domain can be conformally transformed into the interior of a circle. In this transformation R becomes a doubly-connected domain whose frontier consists of a continuum C_1' and a circle C_2'. The simply-connected domain exterior to C_1' and containing C_2' in its interior is now transformed into the interior of the unit-circle. In this transformation C_2' is transformed into an analytic curve. It can therefore be assumed, without loss of generality, that the doubly-connected domain R has as its frontier the unit-circle C_1 and an analytic curve C_2, without double-points, surrounding the point $z = 0$.

By means of the transformation $z = e^u$, a certain periodic curvilinear strip R_u in the u-plane is conformally transformed into the domain R. The set of points corresponding to a given point z_0 is of the form

$$u_k = u_0 + 2ik\pi \quad (k = 0, \pm 1, \pm 2, \ldots).$$

Thus the points equivalent to u_0 are generated by a *cyclic* group. Since R_u is simply-connected, a one-one conformal transformation $u = \phi(w)$ can be found which represents it on $|w| < 1$. The points of $|w| < 1$ which are equivalent to w_0 are obtained as the images of the points u_k. The corresponding group is certainly cyclic and is generated by repetition of a certain non-Euclidean translation, the possibility of a limit-rotation being excluded by the result of § 114.

116. A non-Euclidean translation is a transformation with two fixed points, A and B, on the circumference of the circle. Let

$$w = \psi(t)$$

be a function transforming the interior of the strip $|\Re t| < h$ conformally into the circular area $|w| < 1$ (§ 52), in such a way that the parts of the strip remote from the origin correspond to neighbourhoods of the points A and B respectively. It is easy to see that, in these circumstances, the non-Euclidean translation considered above corresponds to an ordinary Euclidean translation of the strip within itself. The value of h can be so adjusted that the shift in the strip is $2\pi i$.

Let $t = \log \omega$. The function

$$z = e^{\phi(\psi(\log \omega))} = \Phi(\omega)$$

gives a one-one conformal transformation of an annular region into the doubly-connected domain R.

THEOREM 5. *A doubly-connected domain R whose frontier consists of two continua each containing more than one point can be represented, by a one-one conformal transformation, on an annular region whose frontier consists of two concentric circles.*

117. Let R be a simply-connected domain on which the circular area $|w| < 1$ is conformally represented. Two fixed points, z_1 and z_2 of R, correspond to two points w_1 and w_2 which are uniquely determined for any *possible* conformal representation of the circle on R, and whose non-Euclidean distance $D(w_1, w_2)$ is always the same. This number may therefore be regarded as the non-Euclidean distance of the points z_1 and z_2, and in this way a non-Euclidean metric is set up for R.

It is also possible to set up a non-Euclidean metric for a multiply-connected domain R, but the process is somewhat different, for here an infinite number of points w correspond to a single point z. Let γ_z be any rectifiable curve in R. It is transformed into an infinite number of curves $\gamma_{w_1}, \gamma_{w_2}, \dots$ in the w-plane. All these curves can be transformed into one another by non-Euclidean motions, and thus they all have the same non-Euclidean length. This length we associate with the curve γ_z. *In the multiply-connected domain R curves have length, but the conception of a distance between two points has no meaning.*

118. Consider an annular region

$$r_1 < |w| < r_2 \qquad \dots\dots(118\cdot1)$$

and in it a closed curve γ which makes just one circuit round the smaller circle. This curve has non-Euclidean length L_γ. As γ varies let the lower bound of L_γ be α. The number α can be calculated, by the methods of Chapters II and III, and is found to be

$$\alpha = 4\pi^2/\log\,(r_2/r_1).$$

This shows that conformal transformation of one annulus into another is possible if and only if the two annuli are similar figures.

119. Normal families composed of functions which transform simple domains into circles.

Let $\{T\}$ be a family of simple simply-connected domains in the w-plane, each domain containing the point $w = 0$ while none of them contains the point $w = \infty$. A family $\{f(z)\}$ is formed of analytic functions by means of which the domains T are transformed into the circle $|z| < r$. We assume that for all these functions $f(0) = 0$.

Results already obtained show that the functions of the family $\{f(z)\}$ are uniformly bounded in the circle $|z| < r\theta$, where θ is any positive number less than unity, and that in consequence the family is normal in the circle $|z| < r$ if one or other of the following conditions is satisfied:

(α) for each domain T the distance a of the point $w = 0$ from the frontier is not greater than a fixed number M (§ 80 and § 101),

(β) all the functions $f(z)$ satisfy the relation $|f'(0)| \leqslant M$ (§ 81).

The theorem of § 104 shows that the addition of the function $f(z) \equiv 0$ makes these families compact. In case (β) there is no need to add this function if the condition $|f'(0)| \leqslant M$ is replaced by the stricter condition $0 < m \leqslant |f'(0)| \leqslant M$.

120. The kernel of a sequence of domains.

Take a family $\{T\}$ of domains, and, to fix the ideas, suppose it satisfies condition (α) of § 119. In it choose a sequence T_1, T_2, ... of domains such that the corresponding functions $f_1(z), f_2(z),$ ... converge regularly to a function $f_0(z)$ in $|z| < r$.

We assume that $f_0(z)$ does not vanish identically. Then § 104 shows that it provides a conformal transformation of some domain T_0 into $|z| < r$.

Let A_w be a continuum (a closed connex set) which lies within T_0 and contains the point $w = 0$. It will be shown that if n is sufficiently large A_w also lies in T_n.

Denote by A_z the image of A_w, formed by means of the transformation $w = f_0(z)$. Then A_z is a closed set in the circle $|z| < r$ and there is a number $\tau < 1$ such that A_z lies in $|z| < r\tau$.

Either the above statement holds, or there is an infinite sequence of integers, $n_1 < n_2 < ...$, such that, for each n_k, the function

$$w = f_{n_k}(z) \qquad\qquad ......(120{\cdot}1)$$

transforms the circle $|z| \leqslant r\tau$ into a set of points which does not contain the whole continuum A_w. Consequently the w-plane contains at least one point w_{n_k} which is at the same time a point of A_w and the image, by means of (120·1), of a point z_{n_k} of the circle $|z| = r\tau$. The sequence n_k contains a sub-sequence n_k' such that $\lim_{k \to \infty} z_{n'_k}$ exists. Denote the limit by z_0.

Since the sequence $f_1(z), f_2(z),$ ... is continuously convergent

$$f_0(z_0) = \lim_{k \to \infty} f_{n'_k}(z_{n'_k})$$

Also the point $\lim_{k \to \infty} w_{n'_k} = w_0$ belongs to the closed set A_w. Reference to the construction of the circle $|z| = r\tau$ shows that this is impossible.

121. We now assume that T^* is a domain which contains the point $w = 0$ and has the property which has just been proved to hold for T_0, namely, any continuum A_w which lies in T^* and contains the point $w = 0$ lies in every domain T_n from some n onwards. We shall

then show, first that $f_0(z)$ is not a constant, and, secondly, that T^* is contained in T_0.

Let w_0 be an arbitrary point of T^*. Consider a domain A_w, containing the points w_0 and $w = 0$ and such that both A_w and its frontier are contained in T^*. By hypothesis, A_w is covered by T_n whenever n is sufficiently large; for these values of n, the inverse functions $\phi_n(w)$ of $f_n(z)$ are all defined in A_w. These functions are all less than r in absolute value and so form a normal family.

The sequence $\{\phi_n(w)\}$ yields a sub-sequence $\{\phi_{n_k}(w)\}$ of functions which are all defined in A_w and converge continuously to a function $\psi(w)$. By § 104, the function $\psi(w)$ is either a constant, and, in that case, since $\psi(0) = 0$, it vanishes everywhere, or the equation $z = \psi(w)$ gives a conformal transformation of A_w into a domain lying within the domain $|z| < r$.

In both cases the point $z_0 = \psi(w_0)$ lies in this circle. Thus, if

$$z_k = \phi_{n_k}(w_0), \qquad \ldots\ldots(121\cdot1)$$

it follows that $z_k \to z_0$, and, since z_0 is a point within $|z| < r$, $f_{n_k}(z_k) \to f_0(z_0)$ also holds. But, by (121·1),

$$f_{n_k}(z_k) = w_0,$$

from which it follows that

$$w_0 = f_0(z_0).$$

Now w_0 was an arbitrary point of T^*; thus $f_0(z)$ is not constant, and, further, the equation $w = f_0(z)$ gives a conformal transformation of $|z| < r$ into a domain T_0 which must have T^* as a sub-domain.

The domain T_0, obtained from $|z| < r$ by means of the transformation $w = f_0(z)$, where $f_0(z)$ is the boundary function, has the following property: it is the largest domain such that every continuum, containing the point $w = 0$ and contained in the domain, is covered by all the domains T_n from some value of n onwards.

It has incidentally been shown that, if the sequence $f_1(z)$, $f_2(z)$, ... tends to the function which vanishes everywhere, there is no neighbourhood of $w = 0$ which is covered by all the domains T_n from some n onwards.

122. Let an arbitrary sequence T_1, T_2, ... of simply-connected domains each containing $w = 0$ be given. We may suppose, for example, that the domains satisfy condition (α) of § 119. We associate with the sequence a certain set of points K, called the *kernel* of the sequence.

If the sequence $\{T_n\}$ is such that no circle with $w = 0$ as centre is covered by all the domains T_n when n is sufficiently large, then its kernel consists *of the single point* $w = 0$. *In all other cases the kernel K of the sequence is the largest domain having the property that every continuum, containing $w = 0$ and contained in K, lies within every domain T_n for sufficiently large values of n* (22).

The kernel of a sequence of domains is uniquely determined. Suppose that w_0, a point of the w-plane, has a neighbourhood which is covered by all the domains T_n from some n onwards. Let κ_{w_0} be the largest circle with centre w_0 and such that all smaller concentric circles are covered by T_n for sufficiently large values of n. Now let w_0 vary within the w-plane. The sum of the interiors of all the circles κ_{w_0} is either the null-set or an open set which may be regarded as a sum of non-overlapping domains. If one of these domains contains $w = 0$, then that domain is the kernel K. Otherwise the kernel is the single point $w = 0$.

123. *A sequence $\{T_n\}$ of domains is said to converge to its kernel K if every sub-sequence $T_{n_1}, T_{n_2}, \ldots$ has the same kernel as the original sequence $\{T_n\}$.*

Since the domains T_n are assumed to be simply-connected, there are, by § 112, functions $f_1(z), f_2(z), \ldots$ which give conformal transformations of $T_1, T_2, \ldots$ on the circle $|z| < 1$ and satisfy

$$f_n(0) = 0, \quad f_n'(0) > 0. \qquad \ldots\ldots(123\cdot1)$$

Suppose that the functions $f_n(z)$ converge regularly to a function $f_0(z)$ in this circle. Then, if $f_{n_1}(z), f_{n_2}(z), \ldots$ is any sub-sequence,

$$\lim_{k \to \infty} f_{n_k}(z) = f_0(z). \qquad \ldots\ldots(123\cdot2)$$

The function $f_0(z)$ transforms the circle $|z| < 1$ into the kernel K' of the sequence $\{T_{n_k}\}$. From this it follows that $K' = K$, that is, the sequence $\{T_n\}$ converges to its kernel.

The converse of this result also holds: the convergence of a sequence of domains to its kernel implies the regular convergence of the sequence $\{f_n(z)\}$ in the circle $|z| < 1$. For the functions $f_n(z)$ form a normal family. If the sequence $\{f_n(z)\}$ were not convergent it would be possible to find two sub-sequences,

$$f_{n_1}(z), \quad f_{n_2}(z), \ldots, \qquad \ldots\ldots(123\cdot3)$$

$$f_{m_1}(z), \quad f_{m_2}(z), \ldots, \qquad \ldots\ldots(123\cdot4)$$

converging regularly in $|z| < 1$ to two distinct functions $\phi(z)$ and $\psi(z)$ respectively. But this gives a contradiction, for both functions $\phi(z)$ and $\psi(z)$ transform the kernel of the sequence $\{T_n\}$ into the circle $|z| < 1$, and also, from (123·1),

$$\phi(0) = 0, \quad \phi'(0) > 0; \quad \psi(0) = 0, \quad \psi'(0) > 0. \ \ldots\ldots(123\cdot5)$$

The uniqueness theorem for conformal representation (§ 68) shows that $\phi(z) \equiv \psi(z)$.

124. Examples.

(a) Let the w-plane be cut along the negative real axis from the point $-\infty$ to the point $-\dfrac{1}{n}$. Denote by $f_n(z)$ the function which gives a conformal transformation of the circular area $|z| < 1$ into this domain, $f_n(z)$ satisfying the conditions (123·1). Then $f_n(z)$ converges continuously to zero, for the kernel of the sequence of domains is a single point. Actually, we find

$$f_n(z) = \frac{4z}{n(1-z)^2}. \qquad \ldots\ldots(124\cdot1)$$

(b) Let the w-plane be cut along those parts of the imaginary axis which lie above the point i/n and below the point $-i/n$ respectively. Consider the function $f_n(z)$ which transforms the half-plane $\Re(z) > 0$ into this domain and satisfies the conditions

$$f_n(1) = 1, \quad f_n'(1) > 0. \qquad \ldots\ldots(124\cdot2)$$

The general theory shows that $f_n(z)$ converges continuously to z in the half-plane. This can be verified at once; for calculation shows that

$$f_n(z) = \frac{1}{2}\left(z + \frac{1}{z}\right) + \frac{\sqrt{1+n^2}}{2n}\left(z - \frac{1}{z}\right). \qquad \ldots\ldots(124\cdot3)$$

(c) Let the w-plane be cut along an arc of the unit-circle joining the points $e^{i\pi/n}$ and $e^{-i\pi/n}$, and of length $2\pi(1 - 1/n)$. A simply-connected domain is thus formed, with the point $w = \infty$ as an interior point. Let the interior of the unit-circle, $|z| < 1$, be transformed into this domain by means of a function $f_n(z)$ which satisfies the conditions (123·1). The problem is analogous to (b), and it is found that

$$f_n(z) = z\,\frac{1 - z\cos\pi/2n}{\cos\pi/2n - z}. \qquad \ldots\ldots(124\cdot4)$$

As we might anticipate, each of these functions has a pole within $|z| < 1$. However, they form a normal family. This is most readily seen if we observe that our domains can all be transformed into those of example (b) by the use of a *fixed* Möbius transformation.

125. Simultaneous conformal transformation of domains lying each within another.

The solution of the following problem is important for the theorems by which arbitrary analytic functions are represented by single-valued functions (cf. Chap. VIII). Consider an infinite sequence of complex planes, z_1, z_2, ..., and let each plane contain two simply-connected domains, one within the other and both containing the origin.

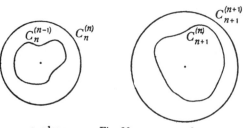

z_n-plane   Fig. 26 z_{n+1}-plane

The two domains in the z_n-plane are denoted by $C_n^{(n)}$ and $C_n^{(n-1)}$ respectively. Take a w-plane and attempt to find in it a sequence $C^{(1)}, C^{(2)}, \ldots$ of domains each lying within the next, all containing the point $w = 0$, and such that the function $\phi^{(n)}(z_n)$, with

$$\phi^{(n)}(0) = 0, \quad \phi'^{(n)}(0) > 0,$$

which transforms $C_n^{(n)}$ into $C^{(n)}$, also transforms $C_n^{(n-1)}$ into $C^{(n-1)}$. The sequence $\{C^{(n)}\}$ is to be such that this holds for all values of n.

Suppose that the domain $C_n^{(n)}$ is transformed conformally into a domain $\Gamma_n^{(n)}$ in the same plane, by means of a function $\psi(z_n)$ with $\psi(0) = 0$ and $\psi'(0) > 0$, and that the same function transforms $C_n^{(n-1)}$ into the domain $\Gamma_n^{(n-1)}$. Then the two original domains may be replaced by their images without materially affecting the problem just stated. From this it follows that there is no loss of generality involved in the assumption that the domains $C_n^{(n)}$ are the circles

$$|z_n| < r_n \quad (n = 1, 2, \ldots). \qquad \ldots\ldots(125{\cdot}1)$$

Let the relation which transforms the circle $|z_n| < r_n$ into the domain $C_{n+1}^{(n)}$ be

$$z_{n+1} = f_{n+1}^{(n)}(z_n), \qquad \ldots\ldots(125{\cdot}2)$$

where it is assumed that $f_{n+1}^{(n)}(0) = 0$ and $f'^{(n)}_{n+1}(0) > 0$. If r_n is fixed, then $f'^{(n)}_{n+1}(0)$ is proportional to r_{n+1}, and r_{n+1} can be determined so as to ensure that

$$f_{n+1}^{(n)}(0) = 0, \quad f'^{(n)}_{n+1}(0) = 1. \qquad \ldots\ldots(125{\cdot}3)$$

We take $r_1 = 1$, and require that conditions (125·3) shall hold for all values of n. Then the numbers r_n are all uniquely determined.

126. The system of equations

$$f^{(n)}_{n+2}(z_n) = f^{(n+1)}_{n+2}\left(f^{(n)}_{n+1}(z_n)\right), \qquad \ldots\ldots(126\cdot1)$$

$$f^{(n)}_{n+k}(z_n) = f^{(n+1)}_{n+k}\left(f^{(n)}_{n+1}(z_n)\right), \qquad \ldots\ldots(126\cdot2)$$

defines new functions $f^{(n)}_{n+k}(z_n)$ which can all be calculated in succession if the functions $f^{(n)}_{n+1}(z_n)$ are known for all values of n. From (125·3) it is seen that

$$f^{(n)}_{n+k}(0) = 0, \quad f'^{(n)}_{n+k}(0) = 1. \qquad \ldots\ldots(126\cdot3)$$

Now observe that the relation

$$w = f^{(n)}_{n+k}(z_n) \qquad \ldots\ldots(126\cdot4)$$

transforms the circle $|z_n| < r_n$ conformally into a *simple* domain of the w-plane. By (126·3), the functions $f^{(n)}_{n+k}(z_n)$, $(k = 1, 2, \ldots, n = 1, 2, \ldots)$, satisfy condition (β) of § 119, so that they form a *normal family*.

127. If n is given, a sequence $k^{(n)}_1$, $k^{(n)}_2$, ... of natural numbers can be found such that the functions

$$f^{(n)}_{n+k^{(n)}_p}(z_n) \quad (p = 1, 2, \ldots) \qquad \ldots\ldots(127\cdot1)$$

form a sequence which converges continuously in the circle $|z_n| < r_n$. Then, by means of the diagonal process, a choice of the numbers $k^{(n)}_p$ can be made which satisfies the condition for all values of n, i.e. a *fixed* sequence of natural numbers $k_1 < k_2 < \ldots$ is found such that

$$\lim_{p \to \infty} f^{(n)}_{n+k_p}(z_n) = \phi^{(n)}(z_n) \quad (n = 1, 2, \ldots), \qquad \ldots\ldots(127\cdot2)$$

within the domain $|z_n| < r_n$.

By § 104 the equations

$$w = \phi^{(n)}(z_n)$$

give conformal transformations of the circle $|z_n| < r_n$ into domains $C^{(n)}$. We shall show that these domains give a solution of our problem. For on writing k_p for k in (126·2) and then proceeding to the limit, as in (127·2), we find

$$\phi^{(n)}(z_n) = \phi^{(n+1)}\left(f^{(n)}_{n+1}(z_n)\right). \qquad \ldots\ldots(127\cdot3)$$

This is exactly what was required.

128. The set of particular domains $C^{(n)}$, which has just been found, forms a figure about which some further observations may be made. From (125·2) using Schwarz's Lemma, it is seen that, if $r_{n+1} \leqslant r_n$, then $|f'^{(n)}_{n+1}(0)| < 1$. Thus (125·3) shows that $r_n < r_{n+1}$ and either

$$\lim_{n \to \infty} r_n = \infty, \qquad \dots\dots(128\cdot1)$$

or there is a finite number R such that

$$\lim_{n \to \infty} r_n = R. \qquad \dots\dots(128\cdot2)$$

These two cases must be considered separately.

129. Denote by a_n the distance of the boundary of $C^{(n)}$ from the point $w = 0$. The function $\phi^{(n)}(z_n)$ transforms the circle $|z_n| < r_n$ into a simple domain, and, by (82·4),

$$a_n > r_n/4. \qquad \dots\dots(129\cdot1)$$

Thus, if (128·1) holds, the sequence $a_1, a_2, \dots$ tends to infinity, so that the sequence of domains $C^{(n)}$ covers the whole w-plane.

130. Now assume that (128·2) holds. The construction shows that $|f^{(n)}_{n+k}(z_n)| < r_{n+k} < R$, so that, on proceeding to the limit,

$$|\phi^{(n)}(z_n)| < R$$

for all values of n. Thus the domains $C^{(1)}$, $C^{(2)}$, $\dots$ lying each within the next, are all contained in the circle $|w| < R$, and the same is true of the set D formed by all points belonging to at least one of them. This set D is itself a simply-connected domain. Let

$$t = \psi(w) \qquad \dots\dots(130\cdot1)$$

be the relation which transforms D into the circular area $|t| < R$, $\psi(w)$ being such that $\psi(0) = 0$, $\psi'(0) > 0$. Schwarz's Lemma shows that

$$\psi'(0) \geqslant 1. \qquad \dots\dots(130\cdot2)$$

On the other hand, the functions

$$t = \Psi_n(z_n) = \psi(\phi^{(n)}(z_n)) \quad (n = 1, 2, \dots) \quad \dots\dots(130\cdot3)$$

give conformal transformations of the circles $|z_n| < r_n$ into domains which lie within $|t| < R$. By Schwarz's Lemma

$$\Psi_n'(0) \leqslant \frac{R}{r_n}. \qquad \dots\dots(130\cdot4)$$

Since $\phi'^{(n)}(0) = 1$, (130·3) shows that

$$\Psi_n'(0) = \psi'(0), \qquad \dots\dots(130\cdot5)$$

so that $\psi'(0) \leqslant R/r_n$. From (128·2) it follows that $\psi'(0) \leqslant 1$, and in (130·2) the sign of equality holds. But this can only be the case if D coincides with the circular area $|w| < R$, i.e. if the domains $C^{(n)}$ tend *uniformly* to the domain $|w| < R$.

CHAPTER VI

TRANSFORMATION OF THE FRONTIER

131. An inequality due to Lindelöf.

Let R be an arbitrary domain in the z-plane, containing the point z_0. The domain R may be multiply-connected. Suppose that there is an arc of the circle $|z - z_0| = r$, subtending an angle

$$\alpha > \frac{2\pi}{n}$$

at the centre z_0 and lying outside the domain R. Here n is a positive integer. By means of rotations about z_0 through angles

$$\frac{2\pi}{n}, \ \frac{4\pi}{n}, \ \dots, \ \frac{2(n-1)\pi}{n},$$

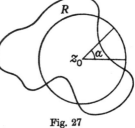

the domain R gives rise to new domains $R_1, R_2, \dots, R_{n-1}$ respectively. The common part $RR_1R_2 \dots R_{n-1}$ of all the domains is an open set which contains the point z_0 but which has no point of the circle

Fig. 27

$|z - z_0| = r$ as an interior point or as a frontier-point. Among the domains whose sum makes up this common part there is one, R_0, which contains the point z_0. The frontier γ of R_0 lies in the circular area $|z - z_0| < r$, and every point of γ is a frontier-point of at least one of the domains $R, R_1, R_2, \dots, R_{n-1}$.

132.
Let $f(z)$ be a bounded analytic function defined in R, and suppose

$$|f(z)| < M \qquad \dots\dots(132\cdot1)$$

in R. Assume further that if ζ is an arbitrary frontier-point of R, lying within the domain $|z - z_0| < r$, and if $z_1, z_2, \dots$ is any sequence of points of R tending to ζ, then

$$\varlimsup_{n \to \infty} |f(z_n)| \leqslant m, \quad (m < M), \qquad \dots\dots(132\cdot2)$$

where m is independent of the particular sequence $\{z_n\}$ and of the choice of ζ. Consider the functions

$$f_k(z) = f\left(z_0 + e^{\frac{2k\pi i}{n}}(z - z_0)\right), \quad \left(k = 1, 2, \dots, (n-1)\right). \qquad \dots\dots(132\cdot3)$$

The function $f_k(z)$ is analytic in the domain R_k. It follows that the function

$$F(z) = f(z)\, f_1(z)\, \dots f_{n-1}(z) \qquad \dots\dots(132\cdot4)$$

is analytic in R_0. Thus, if $\{z_n\}$ is a sequence of points of R_0 tending to a frontier-point ζ as limit, our hypotheses show that

$$\varlimsup_{n\to\infty} |\,F(z_n)\,| \leqslant M^{n-1}m, \qquad \dots\dots(132\cdot5)$$

and hence, since $(132\cdot5)$ holds for all frontier-points of R_0,

$$|\,F(z_0)\,| \leqslant M^n\,\frac{m}{M}.$$

By $(132\cdot3)$ and $(132\cdot4)$,

$$F(z_0) = (f(z_0))^n,$$

so that, finally,

$$|f(z_0)| \leqslant M\left(\frac{m}{M}\right)^{\frac{1}{n}}. \qquad \dots\dots(132\cdot6)$$

133. The following theorem follows almost immediately from Lindelöf's inequality $(132\cdot6)$:

THEOREM. *If $f(z)$ is bounded and analytic in an arbitrary simply-connected domain R whose frontier contains more than one point, and if there is a frontier-point ζ with a neighbourhood N_ζ such that $f(z)$ is continuous at those points and frontier-points of R which lie within N_ζ, and takes the constant value α at those frontier-points, then $f(z) \equiv \alpha$.*

First suppose that ζ is a limiting point of points which do not belong to R or to its frontier. Then, if z_0 is any point of R sufficiently near ζ, we can construct a circle $|z - z_0| \leqslant r$ lying entirely within N_ζ and having points which do not belong to R or its frontier on its circumference. On applying Lindelöf's result to the function $(f(z) - \alpha)$ we obtain $f(z_0) = \alpha$, and the required result follows at once.

If ζ does not fulfil the above condition, we consider the function $F(t) = f(\zeta + t^2) - \alpha$. Comparison with a similar substitution used in § 65 shows that this function is defined in domains R_t which have the property that $t = 0$ is a limiting point of points exterior to R_t. The argument used above therefore serves for this case.

134. Lemma 1, on representation of the frontier.

Let a bounded simply-connected domain R_w be given in the w-plane and suppose that the function

$$w = f(z) \qquad \dots\dots(134\cdot1)$$

represents it conformally on the domain R_z in the z-plane. We assume that the frontier of R_z is a closed Jordan curve c, and that the origins O_w and O_z lie within the respective domains R_w and R_z, and are corresponding points in the transformation.

Let γ_w be a cut into the interior of the domain R_w. By this is meant a Jordan curve joining an interior point of R_w to a point ω of the frontier, and such that any infinite sequence of points of γ_w either has at least one limiting point within R_w or else has ω as its only limiting point.

We shall prove that: *The cut γ_w is transformed into a curve γ_z which is a cut into the interior of the domain R_z bounded by the Jordan curve c.*

135. To prove this lemma we use Jordan's Theorem, which states that c divides the plane into two and only two domains, and also the simplest properties of the domain R_z. These are, that every point of the Jordan curve c is a limiting point of points exterior to R_z, and that every point of c can be joined to any interior point by means of a cut into the interior.

Suppose, if possible, that sequences of points of γ_z can be found, which have two distinct points M and N of c as limiting points. It will be shown that this leads to a contradiction.

The two points M and N divide the Jordan curve c into two arcs, c_1 and c_2. Imagine the curve c described in the positive sense and choose, in order, four points A, B, C and D, of which the first two lie on c_1 and the others on c_2. Let the four points be joined to O_z by four cuts without other intersections than that at O_z. These cuts divide R_z into four sub-domains. By hypothesis, we can find on γ_w a sequence of points which converges to ω, but such that the corresponding points in the z-plane lie alternately in the domain $O_z DA$ and in the domain $O_z BC$. An arc of γ_z joining two successive points of the transformed sequence must cross one or other of the

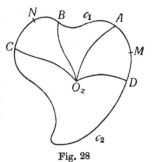

Fig. 28

domains $O_z AB$ and $O_z CD$. To fix the ideas suppose that there is an infinite number of these arcs crossing $O_z AB$, and let the parts of the arcs which lie in this domain and have their end points on $O_z A$ and $O_z B$ be $\Delta_z^{(1)}, \Delta_z^{(2)}, \dots$. The corresponding arcs $\Delta_w^{(1)}, \Delta_w^{(2)}, \dots$ on γ_w are all distinct and tend uniformly to ω. The arcs $\Delta_z^{(1)}$, $\Delta_z^{(2)}$, ... tend uniformly to the arc AB of c. Consider the domain $O_z AB$. On AB choose a point E whose distance from the arcs BO_z and $O_z A$ may be denoted by $2\eta > 0$. In the domain $O_z AB$ take a point z_0 whose distance from E is less than η, and with centre z_0 draw a circle K of radius η. The point E

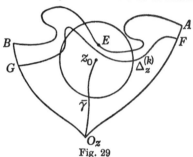

Fig. 29

lies within K and is a limiting point of points exterior to R_z, so that there is an arc K_1 of K lying outside R_z. Let α be the angle subtended at z_0 by this arc, and choose a positive integer n such that $\alpha > 2\pi/n$. Join O_z to z_0 by means of an arc $\bar{\gamma}$ lying within $O_z AB$. If k is sufficiently large, say $k > k_0$, $\Delta_z^{(k)}$ has no points in common with $\bar{\gamma}$, so that the domain $O_z F \Delta_z^{(k)} G$ (Fig. 29) is a sub-domain of $O_z AB$ containing z_0 but not containing any point of K_1 either in its interior or on its frontier.

Let ϵ be an arbitrary positive number. Denote by M the (finite) diameter of R_w and choose k greater than k_0 and also such that the arc $\Delta_w^{(k)}$ lies entirely within the circle

$$|w - \omega| < M\epsilon^n.$$

We now apply Lindelöf's inequality to the function $(f(z) - \omega)$, considered in the domain $O_z F \Delta_z^{(k)} G$. The points of the frontier of this domain which lie within K are all points of $\Delta_z^{(k)}$, so that we may write

$$m = M\epsilon^n.$$

The relation (132·6) shows that $|f(z_0) - \omega| < M\epsilon$. Since ϵ was arbitrary,

$$f(z_0) = \omega.$$

This gives the desired contradiction; for the point $w_0 = f(z_0)$ lies within R_w, whereas ω is a frontier-point of this domain.

136. Lemma 2.

Consider two cuts γ_w' and γ_w'' into the interior of the domain R_w. We suppose that they join the point O_w to the points ω' and ω'' of the frontier and have only the point O_w in common. The result just proved shows that $\gamma_w{}'$ and $\gamma_w{}''$ are the images of two cuts into the interior of R_z. These cuts $\gamma_z{}'$ and $\gamma_z{}''$ join the point O_z to the points ζ' and ζ'' of the Jordan curve c. We now investigate *necessary and sufficient conditions that ζ' and ζ'' should coincide.*

The curve $(\gamma_z{}' + \gamma_z{}'')$ divides R_z into two domains $R_z^{(1)}$ and $R_z^{(2)}$. If ζ' and ζ'' coincide, then, since c is a Jordan curve, one of the domains, say $R_z^{(1)}$, is such that all its frontier-points are points of $\gamma_z{}'$ or of $\gamma_z{}''$ or of both. This domain corresponds to a domain, $R_w^{(1)}$ say, which may have points of the frontier of R_w as frontier-points, but such a frontier-point ω of $R_w^{(1)}$ cannot be at a positive distance from the curve $(\gamma_w{}' + \gamma_w{}'')$. For suppose that ω is such a point. There is a neighbourhood N_ω of ω such that if $w_1, w_2, \ldots$ is a sequence of points converging to a point of the frontier of $R_\omega^{(1)}$ within N_ω, then the inverse of the function $f(z)$, say $\phi(w)$, takes values converging to ζ. By § 133, $\phi(w)$ is constant; but this is impossible.

On the other hand, if $R_w^{(1)}$ has the property in question, so that, in particular, $\omega' = \omega''$, then $\zeta' = \zeta''$. This is readily proved if the above argument is applied to the transformation $w = f(z)$ and the domain $R_z^{(1)}$.

137. Transformation of one Jordan domain into another (23).

We now assume that the frontiers of both the domains R_w and R_z are Jordan curves, which may be denoted by c_w and c_z. Let ω be an arbitrary point of c_w and let $w_1, w_2, \ldots$ denote an arbitrary sequence of points of R_w tending to ω. The corresponding points of R_z are $z_1, z_2, \ldots$.

Since c_w is a Jordan curve the points $w_1, w_2, \ldots$ may be joined, each to its successor, by a sequence of arcs whose aggregate forms a cut γ_w from ω into the interior of R_w.

By § 135, γ_w is the image of a cut γ_z from some point ζ of c_z into the interior of R_z. Also, γ_z contains the sequence of points $z_1, z_2, \ldots$, so that this sequence must converge to ζ. The convergence of the sequence $\{z_n\}$ to ζ is simply a consequence of the convergence of $\{w_n\}$ to ω. It follows that the point ζ to which $\{\zeta_n\}$ converges depends only on ω, not on the choice of the sequence $\{w_n\}$. A point ζ corresponds uniquely to each point ω, and, since R_z and R_w may be interchanged without modification of the reasoning, a point ω of c_w corresponds uniquely to each point ζ of c_z.

It is seen at once that two distinct points ω_1 and ω_2 of c_w have two distinct corresponding points ζ_1 and ζ_2 of c_z. (This also follows from

§ 136.) Further, the transformation of one frontier into the other is continuous. For let $\omega_1, \omega_2, \ldots$ be points of c_w converging to ω_0, and let $\zeta_1, \zeta_2, \ldots$ be the corresponding points of c_z.

If k is a positive integer, a point w_k can be found in R_w such that the relations

$$| z_k - \zeta_k | < \frac{1}{k}, \quad | w_k - \omega_k | < \frac{1}{k}$$

are satisfied simultaneously by w_k and its image z_k. But the points $w_1, w_2, \ldots$ converge to ω_0. It follows that $z_1, z_2, \ldots$, and therefore also the points $\zeta_1, \zeta_2, \ldots$, tend to a point ζ_0.

Finally, let ω_1, ω_2 and ω_3 be three points of c_w which are passed through in this order when the curve is described in the positive sense. Suppose that three cuts into the interior of R_w join O_w to the respective points ω_1, ω_2 and ω_3, and that these cuts do not intersect one another except at O_w. We consider the corresponding figure in the z-plane and observe that the transformation is conformal at O_w. Then it is clear that the points ζ_1, ζ_2 and ζ_3 occur on c_z in this order when the curve is described in the positive sense. These results may be summarized as follows:

THEOREM. *If one Jordan domain is transformed conformally into another, then the transformation is one-one and continuous in the closed domain, and the two frontiers are described in the same sense by a moving point on one and the corresponding point on the other.*

138. A slight generalization of this theorem is easily made. Let R_w be a domain, and c_w a Jordan curve (with or without its end points), and suppose that the following conditions are satisfied: (a) every point ω of c_w is a frontier-point of R_w, (b) every point ω of c_w can be joined to any interior point O_w by a cut into the interior of R_w, (c) every Jordan domain whose frontier consists of a portion $\omega_1\omega_2$ of c_w and two cuts into the interior, $O_w\omega_1$ and $O_w\omega_2$, lies entirely within R_w. Then R_w is said to contain a *free Jordan curve*.

Suppose that $\omega_1 \neq \omega_2$. Then § 136 shows that the cuts $O_w\omega_1$ and $O_w\omega_2$ are the images of two cuts $O_z\zeta_1$ and $O_z\zeta_2$, where $\zeta_1 \neq \zeta_2$. The transformation $w = f(z)$ then transforms the interior of the Jordan domain $O_z\zeta_1\zeta_2O_z$ into the interior of the Jordan domain $O_w\omega_1\omega_2O_w$. By § 137, any arc of the free Jordan curve c_w, and hence the whole curve, is a one-one continuous image of an arc of the frontier c_z of R_z.

Just as in § 136 it may be shown that c_w is not the image of the whole frontier of R_z except when c_w is the whole frontier of R_w, i.e. when R_w is a Jordan domain.

139. Inversion with respect to an analytic curve.

A real analytic curve. in the xy-plane is given either by an equation

$$F(x, y) = 0 \qquad \ldots\ldots(139\text{·}1)$$

or in parametric form by two equations

$$x = \phi(t), \quad y = \psi(t). \qquad \ldots\ldots(139\text{·}2)$$

Here it is supposed that if the functions $F(x, y)$, $\phi(t)$ and $\psi(t)$ are expanded in power-series in the neighbourhood of any point of the curve the coefficients in the expansions are real. Further, we suppose that at each such point at least one of the partial derivatives F_x and F_y and at least one of the derivatives $\phi'(t)$ and $\psi'(t)$ differs from zero.

Consider the function

$$z = f(t) = \phi(t) + i\psi(t) \qquad \ldots\ldots(139\text{·}3)$$

as an analytic function of the complex variable t in the neighbourhood of a real point t_0. By hypothesis $f'(t_0) \neq 0$, so that a circle $|t - t_0| < r$ can be chosen which is conformally represented, by means of (139·3), on a certain neighbourhood of the point $z_0 = f(t_0)$. The circle has a diameter lying on the real axis, and this diameter is transformed into part of the curve (139·2). The following definition is due to Schwarz: Two points z and z^* are said to be *inverse points* with respect to the curve (139·2), if they are the images of two points t and $\bar{t}$ of the circle $|t - t_0| < r$, where t and $\bar{t}$ are conjugate complex numbers. Thus, to (139·3) we can add the equation

$$z^* = \phi(\bar{t}) + i\psi(\bar{t}). \qquad \ldots\ldots(139\text{·}4)$$

Let the complex number conjugate to z^* be denoted by w. Since $\phi(t)$ and $\psi(t)$ can be expanded in the form of power-series with real coefficients, $\phi(\bar{t})$ and $\phi(t)$, $\psi(\bar{t})$ and $\psi(t)$ form pairs of conjugate numbers. Hence

$$w = \bar{z}^* = \phi(t) - i\psi(t); \qquad \ldots\ldots(139\text{·}5)$$

(139·3) and (139·5) give

$$x = \phi(t) = \frac{z + w}{2}, \quad y = \psi(t) = \frac{z - w}{2i}. \qquad \ldots\ldots(139\text{·}6)$$

Substitution of these values in (139·1) gives

$$F\left(\frac{z + w}{2}, \ \frac{z - w}{2i}\right) = 0, \qquad \ldots\ldots(139\text{·}7)$$

and from this relation w can be calculated as an analytic function of z. The relation (139·7) shows, in particular, that the operation of inversion depends only on the form of the curve (139·1), and is independent of the choice of the parameter t in (139·2) (24).

140. The equation (139·7) is especially convenient when an algebraic curve is given. If (139·1) represents a straight line or a circle, then the corresponding formulae (139·7) are exactly the inversion formulae given in Chapter I.

In the case of the ellipse

$$\frac{x^2}{a^2} + \frac{y^2}{b^2} = 1, \qquad \text{......(140·1)}$$

(139·7) becomes

$$(a^2 - b^2)(w^2 + z^2) - 2(a^2 + b^2) zw + 4a^2 b^2 = 0. \quad \text{......(140·2)}$$

If this is solved for w a two-valued function of z is obtained. This function is regular throughout the z-plane except at the foci of the ellipse (140·1). The corresponding transformation

$$(a^2 - b^2)(z^{*2} + \bar{z}^2) - 2(a^2 + b^2) z^* \bar{z} + 4a^2 b^2 = 0 \quad \text{......(140·3)}$$

can only be regarded as an inversion with respect to the ellipse when the points transformed lie within the ellipse confocal to (140·1) and passing through the point

$$x_1 = \frac{a^2 + b^2}{\sqrt{a^2 - b^2}}, \quad y_1 = 0.$$

141. The inversion principle.

Let R_v be a simply-connected domain in the v-plane, containing a segment $A_1 B_1$ of the real axis, and symmetrical with respect to this axis. Suppose that the relation $v = \psi(t)$ gives a conformal transforma-

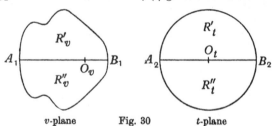

v-plane   Fig. 30   t-plane

tion of R_v into the interior of the unit-circle in the t-plane, the transformation being such that the origin $t = 0$ corresponds to some point O_v of $A_1 B_1$, and that $\psi'(0) > 0$. The function $\psi(t)$ is uniquely determined by these conditions (§ 112). Further, if the two figures are inverted with respect to $A_1 B_1$ and $A_2 B_2$ respectively, they are transformed into themselves. From this it follows that

$$\bar{\psi}(\bar{t}) = \psi(t), \qquad \text{......(141·1)}$$

where $\bar{\psi}$ and $\bar{t}$ are the numbers conjugate to ψ and t.

The relation (141·1) shows that $\psi(t)$ may be written in the form of a power-series with *real* coefficients, so that the segments $A_1 B_1$ and $A_2 B_2$ correspond to one another. Hence R_v' and R_t' are corresponding domains.

142. To obtain a conformal transformation of R_v' into the interior of a unit-circle $|z| < 1$, we may first use the relation $v = \psi(t)$ to transform R_v' into R_t' and then a relation $t = \phi(z)$ to transform R_t' into $|z| < 1$.

The function $\phi(z)$ is already known (§ 54); it transforms the circular area $|t| < 1$ into the z-plane cut along an arc of $|z| = 1$. Two points, z_1 and z_2, which are inverse points with respect to $|z| = 1$, correspond to two points which are symmetrical with respect to the real axis in the t-plane.

The transformation $v = \psi(\phi(z))$ therefore transforms $|z| < 1$ into R_v' in such a way that the points z_1 and z_2 correspond to points v_1 and v_2 which are symmetrically placed with respect to $A_1 B_1$. The segment $A_1 B_1$ is transformed into an arc of the circle $|z| = 1$, and the function $\psi(\phi(z))$ is analytic on this arc.

143. Let c_w be a regular analytic curve, and suppose that the relation

$$w = \chi(v) \qquad \ldots\ldots(143\cdot1)$$

transforms it into a segment of the real axis in the v-plane. Let a domain R_w' in the w-plane, having an arc of c_w as a free Jordan curve, be trans-

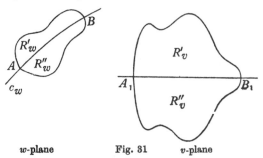

w-plane   Fig. 31   v-plane

formed by (143·1) into a domain R_v'. Let the domain R_w'', which is obtained by inversion of R_w' with respect to c_w, be transformed conformally, by means of (143·1), into the inverse R_v'' of R_v'.

In order to obtain a conformal transformation of R_w' into the circle $|z| < 1$, it is only necessary to substitute the function $v = \psi(\phi(z))$ of § 142 in equation (143·1). We obtain the transformation

$$w = f(z), \qquad \ldots\ldots(143\cdot2)$$

which has the following properties: the arc AB of c_w, which is part of the frontier of R_w', is the image of an arc of $|z| = 1$ and on this arc $f(z)$ is an analytic function; if z_1 and z_2 are inverse points with respect to $|z| = 1$, the corresponding points w_1 and w_2 lie one in R_w' and the other in R_w'', and w_1 and w_2 are inverse points with respect to c_w.

144. Let R_w' be the domain just considered, and let R_v' be a domain lying within some circle and having an arc A_3B_3 of the circle as part of its frontier. Suppose that the relation $w = F(v)$ gives a conformal transformation of R_v' into R_w' in which the circular arc A_3B_3 corresponds to the arc AB of c_w. Then $F(v)$ has, on A_3B_3, exactly the same properties as were found to hold for $f(z)$ in § 143. For $F(v)$ is obtained by elimination of z between the equations $w = f(z)$ and $v = g(z)$, the latter being a relation by which R_v' is transformed into $|z| < 1$.

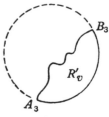

Fig. 32

Now let R_w^0 be an arbitrary simply-connected domain in the w-plane, having an arc MN of an analytic curve c_w as a free Jordan curve. Let

$$w = F(v) \qquad \ldots\ldots(144\cdot1)$$

transform R_w^0 into the circular area $|v| < 1$. In general it is not possible to invert the *whole* domain R_w^0 with respect to c_w, so that the method of § 143 cannot be used. We know, however, that the arc MN of c_w is transformed into an arc of the circle $|v| = 1$, and that this is a one-one continuous transformation (§ 138). Now let ω be an arbitrary point of MN. We

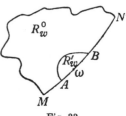

Fig. 33

can find a sub-domain R_w' of R_w^0, which can be inverted with respect to c_w and which has, as part of its frontier, an arc AB of c_w, where AB has ω as an interior point and is itself part of MN. The relation $(144\cdot1)$ transforms R_w' into a domain R_v' which has all the properties mentioned at the beginning of this paragraph.

Thus $F(v)$ is analytic at all points of the arc of $|v| = 1$ which corresponds to c_w. Further, every interior point of this arc has a neighbourhood such that a pair of inverse points v_1 and v_2, both lying in the neighbourhood, are transformed into a pair of inverse points with respect to c_w.

These facts constitute the famous Principle of Symmetry, or Inversion Principle, associated with the name of Schwarz (25).

145. Let R be an arbitrary simply-connected domain whose frontier consists of more than one point. Suppose that the function

$$w = f(z) \qquad \qquad \text{......(145·1)}$$

gives a conformal transformation of the circle $|z| < 1$ into the domain R. As in Chapter II, we now regard the circle as a non-Euclidean plane. The non-Euclidean straight lines in $|z| < 1$ are such that if the circular area is inverted with respect to one of them it is transformed into itself. An analytic curve in the w-plane which corresponds to one of these non-Euclidean straight lines divides R into two sub-domains R' and R'', and these are interchanged when they are inverted with respect to the curve. Thus all these curves may be regarded as lines of symmetry in R.

146. Transformation of corners.

Let R_w be a domain whose frontier contains the free Jordan curve MN, and let A be an interior point of MN at which the portions AM and AN of the curve both possess tangents, which we denote by AP and AQ respectively. Then R_w has a corner at A, and this is measured by the angle α between the two directions AP and AQ.

If R_w is transformed into the circular area $|z| < 1$, the point A corresponds to a certain point A_1 of the circumference $|z| = 1$. It will now be proved that any curve in R_w which ends at A and has a tangent at A making an angle $\theta\alpha$ with AQ, $(0 < \theta < 1)$, is the image of a curve in the z-plane, this curve having at A_1 a tangent which makes an angle $\theta\pi$ with the circular arc $A_1 N_1$.

It is easy to deal with the case where AM and AN are both straight lines. For a circular sector S_w can be found, having its straight sides on

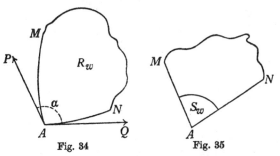

Fig. 34   Fig. 35

AM and AN and lying entirely within R_w. The function $z = \phi(w)$, which transforms R_w into $|z| < 1$, transforms S_w into a sub-domain S_z of the unit-circle, and this sub-domain has A_1 as a frontier-point. It is only necessary to show that angles at A are transformed proportionally into angles at A_1.

Now if A is the point w_0, the transformation $u = (w - w_0)^{\frac{\pi}{a}}$ represents S_w on a semi-circle S_u, and S_u can then be transformed into S_z. But the ratio of the angles in question is clearly preserved in the transformation of S_w into S_u, and the second transformation, S_u into S_z, is analytic (§ 143) and therefore conformal.

147. Our proof of the general theorem depends upon the following lemma:

Let the domain R contain the domain S, but suppose that R and S have the free Jordan curve ACB as part of their respective frontiers. Let R and S separately be transformed into the same circular area K in such a way that in both cases the ends A and B of the Jordan curve

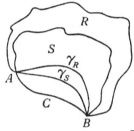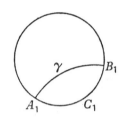

Fig. 36

ACB are transformed into two fixed points A_1 and B_1 of the circumference of K. Now let γ be a circular arc within K, joining A_1 and B_1, and let γ_R and γ_S be its images. It will be shown that γ_S lies within the domain whose frontier consists of ACB and γ_R.

If, by a conformal transformation, the circle K is transformed into a semi-circle, the arc $A_1C_1B_1$ being made to correspond to the diameter A_2B_2 of the semi-circle, then the curve γ is transformed into a circular arc passing through A_2 and B_2. We may assume that A_2B_2 lies on the real axis. From this it is seen that a transformation of R into a semi-circle, with correspondence

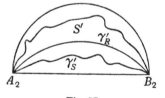

Fig. 37

between the Jordan curve ACB and the diameter, transforms γ_R into a circular arc γ_R' passing through A_2 and B_2. The same transformation represents S on a domain S' lying within the semi-circle, while γ_S is transformed into a curve γ_S' within the semi-circle and joining A_2 and B_2. We have only to show that γ_S' lies between γ_R' and A_2B_2.

Now γ_S' may be regarded as the curve obtained from γ_R' by means of the transformation which represents the semi-circle conformally on S'

and transforms the diameter $A_2 B_2$ into itself. The inversion principle shows that the function which sets up this transformation is analytic throughout the circle whose diameter is $A_2 B_2$. The function takes real values on $A_2 B_2$. It is now seen, as a direct consequence of the theorem of § 88, that γ_s' lies between γ_R'' and $A_2 B_2$.

148. We are now in a position to prove the theorem of § 146 in the case where R has a corner A, one of whose arms, say AN, is a straight line.

As before, let α denote the measure of the corner at A. Now construct two new domains, R' and R'', as follows: R' is a domain containing R. Its frontier contains AN and also a segment AM' which makes an angle $(\alpha + \epsilon)$ with AN. The domain R'' is contained in R. Its frontier contains AN and also a segment AM'' which makes an angle $(\alpha - \epsilon)$ with AN.

Take three functions by means of which the three domains R, R' and R'' are transformed into a fixed circle, in such a way that the segment AN corresponds to the same circular arc $A_1 N_1$ in each case. Let γ_0 be a circular arc joining A_1 and N_1, lying within the circle and making an angle $\pi \theta$ with the arc $A_1 N_1$. Consider the images

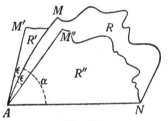

Fig. 38

γ, γ' and γ'' of this arc in the three domains R, R', R''. By § 147, γ lies between γ' and γ''; § 146 shows that the curves γ' and γ'' both have tangents at A, these tangents making angles $\theta(\alpha + \epsilon)$ and $\theta(\alpha - \epsilon)$ respectively with AN. Since ϵ is arbitrary and γ always lies between γ' and γ'', it is seen that γ itself has a tangent at A and that this tangent makes an angle $\theta \alpha$ with AN; for the contrary assumption would lead to a contradiction.

This proves the theorem of § 146 for domains which have a straight line as one arm of the corner A.

149. The general case of § 146 can be made to depend on the result just obtained. Let R be the domain with a corner at A. Consider a domain R' which contains R and has a corner at A, one of whose arms is AM and the other a straight line AK. First transform R' into a half-plane; at the same time R is transformed into a domain R_1 with a corner at A_1, the arm $A_1 M_1$ which corresponds to AM being now a straight line. In this transformation of R into R_1, and a further transformation of R_1 into a circle, the ratio of angles at A is unaltered, as is shown by § 148.

Fig. 39

150. In particular, if the frontier of the domain R has a tangent at A, so that A is a corner with $\alpha = \pi$, then the transformation is isogonal at A. If $f(z)$ is the function which gives the transformation, and if ζ is the point of the circumference corresponding to A, then isogonality is expressed analytically by the statement that the function

$$\arg \frac{f(\zeta) - f(z)}{\zeta - z} \qquad \qquad \ldots\ldots(150{\cdot}1)$$

tends to a constant as z tends to ζ. An analytical proof of isogonality at A has been given by Lindelöf, who used (150·1) and Poisson's integral (23). He further showed that if the curve MAN is smooth (*glatt*) at A, then the function $\arg f'(z)$ is continuous at ζ. (A curve which has a tangent at A is said to be smooth at A if every chord BC tends to the tangent as B and C tend to A simultaneously*.)

It should be noticed that the function $f'(z)$ is not necessarily finite in the neighbourhood of ζ even in the case considered by Lindelöf. For instance, the function

$$w = -z \log z, \qquad \qquad \ldots\ldots(150{\cdot}2)$$

where $z = re^{i\phi}$, transforms the domain lying between the imaginary axis $\Re z = 0$ and the curve

$$\log r = -\phi \frac{\cos \phi}{\sin \phi}, \quad \left(0 < \phi < \frac{\pi}{2} \text{ and } -\frac{\pi}{2} < \phi < 0\right),$$

into the domain lying between the convex curve

$$w = \frac{\pi}{2} r \mp i r \log r, \quad (0 < r \leqslant 1),$$

and the real w-axis cut between the points $w = \dfrac{1}{e}$ and $w = \dfrac{\pi}{2}$. But the derivative of (150·2) is

$$w' = -(1 + \log z), \qquad \qquad \ldots\ldots(150{\cdot}3)$$

and tends to ∞ as z tends to zero; $\arg w'$, however, tends to zero.

151. The theorem of §93 may be completed and can then be applied in many cases, and leads to a more precise result than that just given.

---

* The curve $y = f(x)$ with $f(0) = 0$, $f(x) = x^2 \sin \dfrac{1}{x}$ for $x \neq 0$, has $y = 0$ as tangent at the point $x = 0$, but is not smooth at this point.

Let $f(z)$ be a function regular in $|z| < 1$ and such that $|f(z)| < 1$ in this domain. Consider the function $f(z)$ at points within the triangle ABC, where A is the point $z = 1$, BC is perpendicular to the real axis, lies entirely within the unit-circle, and is at a distance h from A. There is a positive number M such that, for all values of z within the triangle ABC,

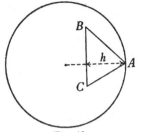

$$|1 - z| < M(1 - |z|), \quad ...(151\cdot1)$$

and hence

$$\frac{1 - |f(z)|}{1 - |z|} < M \left| \frac{1 - f(z)}{1 - z} \right|. \qquad ......(151\cdot2)$$

Fig. 40

Thus either

$$\frac{1 - f(z)}{1 - z}$$

tends uniformly to infinity as z tends to A by values within the triangle, or there is a sequence of points $z_1, z_2, ...,$ lying within the triangle and tending to $z = 1$, such that $\left| \dfrac{1 - f(z)}{1 - z} \right|$ and, by $(151\cdot2)$,

$$\frac{1 - |f(z_n)|}{1 - |z_n|}$$

are bounded; the theorem of § 93 can then be used.

In the latter case, let $z_1, z_2, ...$ be an arbitrary sequence of points within ABC tending to $z = 1$. Let two sequences of numbers, $r_1, r_2, ...$ and $\rho_1, \rho_2, ...,$ be defined by the relations

$$r_n = \frac{1 - \Re z_n}{h}, \quad \rho_n = \frac{\alpha r_n}{1 - r_n(1 - \alpha)}, \qquad ......(151\cdot3)$$

where α has the same meaning as in $(93\cdot10)$. Further, let

$$1 - z = r_n(1 - t), \qquad ......(151\cdot4)$$

$$1 - f(z) = \rho_n(1 - \phi_n(t)). \qquad ......(151\cdot5)$$

The functions $\phi_n(t)$ form a sequence each of whose members is regular in $|t| < 1$, and Julia's Theorem (§ 90) shows that at all points of this circle, and for $n = 1, 2, ...,$ $|\phi_n(t)| < 1$.

By $(151\cdot4)$ z is real when t is real. If t is fixed, $z \to 1$ as $n \to \infty$. By $(93\cdot10)$ and $(151\cdot3)$

$$\lim_{n \to \infty} \frac{1 - \phi_n(t)}{1 - t} = \lim_{n \to \infty} \frac{1 - f(z)}{1 - z} \frac{r_n}{\rho_n} = 1.$$

Thus, when t is real, the functions $\phi_n(t)$ satisfy the relation

$$\lim_{n \to \infty} \phi_n(t) = t. \qquad ......(151\cdot6),$$

Also these functions form a normal family within $|t| < 1$. It follows that $(151 \cdot 6)$ holds uniformly in any closed set of points lying within $|t| < 1$.

Now consider the original sequence of points z_1, z_2, ... and define a sequence of numbers t_1, t_2, ... by means of the equations

$$1 - z_n = r_n (1 - t_n); \quad (n = 1, 2, ...); \qquad \ldots \ldots (151 \cdot 7)$$

$(151 \cdot 3)$ shows that t_1, t_2, ... all lie on the base BC of the triangle ABC. Thus, if $\phi_n(t_n) = t_n + \tau_n$, then $\tau_n \to 0$, for BC is a closed set of points within the unit-circle.

The relations $(151 \cdot 4)$ and $(151 \cdot 5)$ yield

$$\frac{1 - f(z_n)}{1 - z_n} = \frac{\rho_n}{r_n} \left(1 - \frac{\tau_n}{1 - t_n} \right),$$

and this gives

$$\lim_{n \to \infty} \frac{1 - f(z_n)}{1 - z_n} = \alpha. \qquad \ldots \ldots (151 \cdot 8)$$

In conclusion, we observe that since $\{\phi_n(t)\}$ is a normal family, so also is $\{\phi_n'(t)\}$, and that $\lim_{n \to \infty} \phi_n'(t_n) = 1$. Now differentiate $(151 \cdot 5)$ with respect to t, and, taking account of $(151 \cdot 4)$, substitute t_n for t. This gives $f'(z_n) r_n = \rho_n \phi_n'(t_n)$. Thus

$$\lim_{n \to \infty} f'(z_n) = \alpha. \qquad \ldots \ldots (151 \cdot 9)$$

We have proved the following theorem:

THEOREM. *If, in $|z| < 1$, the function $f(z)$ is regular and $|f(z)| < 1$, and if z_1, z_2, ... is any sequence of numbers lying within the triangle ABC and tending to $z = 1$, then*

$$\lim_{n \to \infty} \frac{1 - f(z_n)}{1 - z_n}$$

exists. This limit is either $+\infty$ or it is a number $\alpha > 0$. In the latter case

$$\lim_{n \to \infty} f'(z_n) = \alpha.$$

152. Conformal transformation on the frontier.

Let R be a simply-connected domain and P a point of its frontier through which two circles K and K' can be drawn, K' lying entirely outside and K entirely inside R. We now make a conformal transformation of R into the circular area $|z| < 1$. It will be shown that there is a point A on $|z| = 1$ such that, if z approaches A from within a triangle ABC, whose base BC lies within the circle, the corresponding point $f(z)$ in R approaches P, and also the derivative $f'(z)$ tends to a unique, finite, non-zero limit. It is therefore legitimate to speak of a conformal transformation of frontier-points.

By means of a Möbius transformation, let the interior of the circle K' be represented on the exterior of the unit-circle $|w| = 1$, in such a way that P is transformed into the point $w = 1$. Then R is transformed into a domain R_1 lying within $|w| < 1$. The circle K is transformed into a circle K_1 with centre C_1 lying within R_1. We need only show that if the function giving the transformation is suitably chosen the number α of § 151 is finite. Take that transformation $w = f(z)$ which makes $z = 0$ correspond to the centre C_1 of K_1, and let ρ_1 denote the radius of K_1.

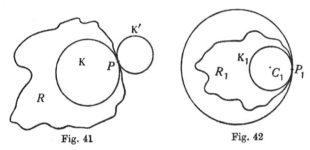

Fig. 41   Fig. 42

By Schwarz's lemma, every circle $|z| = \theta$, $(\theta < 1)$, is transformed into a curve surrounding the circle $|w - (1 - \rho_1)| = \rho_1\theta$. Thus this curve passes through at least one point w' which lies on the real axis and satisfies $1 - w' < \rho_1(1 - \theta)$. If z' is the point corresponding to w', then $(1 - |w'|)/(1 - |z'|) < \rho_1$. We can construct a sequence of points $z_1, z_2, \ldots$ converging to a point ζ of $|z| = 1$ and such that the corresponding points $w_1, w_2, \ldots$ all satisfy the relation $(1 - |w_n|)/(1 - |z_n|) < \rho_1$. Finally, a rotation of the unit-circle ensures that $\zeta = 1$. The conditions of the theorem of § 93 are all satisfied and it is seen that, in the theorem of § 151, $\alpha \leqslant \rho_1$.

153. Attention should be called to the fact that P may be a frontier-point at which the transformation is conformal and yet not lie on a free Jordan curve. For example, consider the unit-circle $|w| < 1$, and in it circular arcs joining -1 to $+1$ and making angles $\pm \dfrac{\pi}{2} \dfrac{n-1}{n}$, $(n = 1, 2, \ldots)$ with the real axis at these points. Now cut the circle along each arc between the point $w = -1$ and the first intersection of the arc with the circle $|w - (1 - \rho_1)| = \rho_1$. The unit-circle, with these cuts, is a domain to which our theorem applies. The function $f(z)$, considered in the whole circle $|z| < 1$, is not even continuous at $z = 1$. But, within an angle at $z = 1$ whose arms are chords of the circle, both $f(z)$ and $f'(z)$ are continuous and bounded (26) (27).

CHAPTER VII

TRANSFORMATION OF CLOSED SURFACES

154. Blending of domains.

In a u-plane let three Jordan arcs a, b and c, all passing through P and Q but with no other common points, define three Jordan domains A, B and $(A + B)$. In a v-plane let three circular arcs b', c' and d', all passing through P' and Q', define two crescents B' and C'; the sum of these domains, $(B' + C')$, is a circle whose circumference is made up of the two arcs b' and d'. Further, the angle between the arcs b' and c' is $\frac{\pi}{2^n}$, where n is a positive integer. Suppose that a known function $u = \psi(v)$ represents B' on B in such a way that b corresponds to b' and c to c'. It will be shown that two functions $z = f(u)$ and $z = g(v)$ can be found, such that the former transforms $(A + B)$ into a domain $(A'' + B'')$ and the latter transforms the circle $(B' + C')$ into a domain $(B'' + C'')$. Here $(A'' + B'' + C'')$ is a circular area. The transformations are to be such that A and A'', C' and C'' are corresponding areas, while B'' corresponds both to B and to B'. Given any point of B'' the corresponding points in the u- and v-planes are to satisfy the relation $u = \psi(v)$.

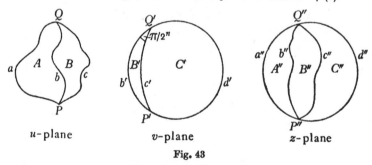

u-plane v-plane z-plane

Fig. 43

First take a function $u_1 = \phi_1(u)$ which transforms $(A + B)$ into the circular area $|u_1| < 1$, in such a way that the centre of the circle is the image of an interior point of A. Let the new domains corresponding to A and B be A_1 and B_1 respectively. The frontier of B_1 consists of a Jordan curve b_1 and a circular arc c_1.

The function $u_1 = \phi_1(\psi(v)) = \psi_1(v)$ transforms the domain B' into B_1 in such a way that the two circular arcs c' and c_1 correspond. The inversion principle shows that $\psi_1(v)$ is defined not in B' alone, but in a crescent which has angles $\pi/2^{n-1}$ at P' and Q'. We may suppose that this crescent is bounded by arcs b' and c'. The function $u = \psi_1(v)$ transforms this new crescent into a domain made up of B_1 and its inverse $\bar{B}_1$ with respect to the circle $a_1 + c_1$. Our problem has thus been reduced to another of the same kind in which the number n has been replaced by $(n-1)$.

By repetition of this process we obtain in succession the functions $u_2 = \phi_2(u_1)$, ..., $z = u_n = \phi_n(u_{n-1})$, and from these $z = f(u)$ can be calculated. It is easy to see that $g(v) = f(\psi(v))$.

155. The special domains B' and C' in the v-plane can now be replaced by much more general ones. We assume that the frontiers of B' and C' consist of three Jordan curves joining the points P' and Q' and having distinct tangents at these points. Let $v = \chi(w)$ be a function which transforms $|w| < 1$ into $(B' + C')$. Then B' is transformed into a domain whose frontier consists of a circular arc β' and a Jordan curve which meets either end of β' at an angle which differs from zero. This shows that the domain just described contains a crescent $\beta'\gamma'$ whose angle is $\pi/2^n$. The function $u = \psi(\chi(w))$ transforms γ' into a curve in the u-plane and, if we cut away the domain that lies between γ and c, our problem is the same as that of § 154.

156. Conformal transformation of a three-dimensional surface (28).

Let a surface S in three-dimensional XYZ-space be represented, in the neighbourhood of some point of S, by means of the equations

$$X = X(\alpha, \beta), \quad Y = Y(\alpha, \beta), \quad Z = Z(\alpha, \beta), \quad \ldots\ldots(156\cdot1)$$

where α and β are parameters. The same surface can be represented in terms of other parameters u and v by substitution in $(156\cdot1)$ of the functions $\alpha = \alpha(u, v)$, $\beta = \beta(u, v)$. We assume that the functions $\alpha(u, v)$ and $\beta(u, v)$ can be so chosen that the new equations for S,

$$X = \xi(u, v), \quad Y = \eta(u, v), \quad Z = \zeta(u, v), \quad \ldots\ldots(156\cdot2)$$

have continuous first partial derivatives in some domain R_w of the uv-plane, and at the same time

$$ds^2 = \lambda(u, v)(du^2 + dv^2), \qquad \ldots\ldots(156\cdot3)$$

where $\lambda \neq 0$ at all points R_w. It then appears that the domain R_w is represented on a portion R_S of S by a one-one continuous correspondence, that any two curves in R_w which have continuously turning tangents and cut at an angle α correspond to two curves which cut at an angle α on R_S, and that the scale of the representation at the point of intersection is independent of direction. In other words the representation of R_w on R_S is *conformal*.

It is known that it is always possible to introduce parameters u and v such that (156·3) holds in a certain neighbourhood of a point (α_0, β_0) of (156·1) provided that in some domain of the $\alpha\beta$-plane the functions $X(\alpha, \beta)$, $Y(\alpha, \beta)$ and $Z(\alpha, \beta)$ have continuous first partial derivatives which satisfy a Lipschitz condition, and that the three Jacobians,

$$\frac{\partial(X, Y)}{\partial(\alpha, \beta)}, \quad \frac{\partial(Y, Z)}{\partial(\alpha, \beta)}, \quad \frac{\partial(Z, X)}{\partial(\alpha, \beta)},$$

do not all vanish at (α_0, β_0). It is conceivable that the equation (156·3) may be satisfied, for a suitable choice of parameters, even on surfaces which do not satisfy all the conditions just mentioned.

157. Conformal representation of a closed surface on a sphere.

Suppose that S is a surface which can, by a one-one continuous transformation, be made to correspond to a closed sphere Σ. Further let every point P of S lie within a portion of S which can be conformally represented on a portion of a plane (§ 156). It will be shown that the whole surface S can be represented conformally on Σ.

Consider an arbitrary one-one continuous transformation of S into Σ. The north pole N_1 of Σ corresponds to a point N of S. Stereographic projection from N_1 represents Σ on a plane T, and there is a one-one continuous correspondence between T and the pricked surface S^* obtained by omitting the point N from S.

158. Now consider in T an infinite sequence of triangles T_1', T_2', ... arranged spirally (like the peel of a peeled apple) and covering the whole plane T. The sum $\sigma_n' = T_1' + T_2' + \ldots + T_n'$ of the first n triangles always covers a simply-connected domain. The figure gives a simple example of what is meant. The triangles may have curvilinear sides; they are drawn with straight sides purely as a matter of convenience. The triangles T_n' are, however, to be so chosen that their

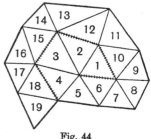

Fig. 44

images T_n on S^* satisfy the two following conditions: (a) the sides of

T_n are curves on S^* which have tangents at their end-points, and two intersecting sides have distinct tangents; (b) any two triangles with a common side lie within a portion of S that can be represented conformally on a plane domain.

159. It will be shown that for $n = 1, 2, \ldots$ the sum $\sigma_n = T_1 + T_2 + \ldots + T_n$ can be represented conformally on a circle $|z_n| < r_n$ in a complex z_n-plane. Assume that σ_{n-1} has already been represented on a circle $|z_{n-1}| < r_{n-1}$, and that the figure $(T_{n-1} + T_n)$ has been represented on a domain of a v-plane. The triangle T_{n-1} appears in both representations, and if the image of a point of T_{n-1} in one is made to correspond to the image of the same point in the other, a new conformal representation is set up. The representation of σ_n on $|z| < r_n$ is thus reduced to the problem which was solved in §§ 154, 155.

160. The transformation of the whole pricked surface S^* has now been reduced to the problem considered in §§ 125–129. The results there obtained show that S^* can be transformed either into the circle $|w| < R$ or into the whole w-plane. We shall show that the second possibility is the true one.

Take a simply-connected neighbourhood of the point N_1 that was removed from S, and transform it conformally into a circle $|u| < 1$. A hypothesis shows that this is possible. The transformation may be so chosen that the pricked circle $0 < |u| < 1$ consists of the images of points of S^*. If now $\omega = 1/w$, there is an analytic function $\omega = \phi(u)$, such that two points u and ω satisfy this relation if they correspond to the same point of S^*. The function $\phi(u)$ is single-valued in the pricked circle $0 < |u| < 1$; it is bounded in the neighbourhood of $u = 0$ and therefore (§ 114) it gives a representation of a pricked domain in the w-plane. But this can only be the case if S^* has been transformed into the whole w-plane.

If we now assume that the w-plane is the stereographic projection of a sphere Σ, the transformation will be conformal even at the north pole of the sphere. The closed surface S has been conformally represented on a closed sphere.

The transformation is determined uniquely if three points are chosen on the surface and are made to correspond to three given points on the sphere.

161. Conformal transformation of polyhedral surfaces into the surface of a sphere was one of the earliest applications of the theory to be attempted. Schwarz showed that both the general tetrahedron and the cube can be transformed into the sphere. The transformation of the most general closed polyhedron into the closed sphere is a problem which is almost identical with the one just treated. We cannot require preservation of angles at edges and corners of the figure, and to specify our intentions we make the following conventions as to representation of neighbourhoods of these singular points.

Along an edge two faces meet. We suppose them rigid, cut them from the figure and spread them out on a plane. Thus a neighbourhood of any point of the edge is represented as a plane neighbourhood.

If C is a corner we make a quasi-conformal representation of a neighbourhood of C onto a plane domain, as follows: By means of a small sphere, centre C, a small neighbourhood of C is cut off. The pyramid-like figure so obtained is cut open along one of its edges and spread out on the u-plane, the point C corresponding to $u = 0$. If now $\pi\alpha$ is the sum of the superficial angles of the polyhedron at C, then the function $v = u^{\frac{2}{\alpha}}$ gives the required quasi-conformal representation of the corner C.

The method used above now gives a solution of our problem (29).

CHAPTER VIII

THE GENERAL THEOREM OF UNIFORMISATION

162. Abstract surfaces.

We take a finite or infinite collection of triangles T_1, T_2, ..., and suppose that with each side of each triangle T_i is associated exactly one side of some other triangle T_k. We may picture associated sides as being welded together; when this has been done, any vertex P will belong to several triangles. We assume that the triangles adjoining a common vertex form a *finite cycle* of adjacent triangles.

The welding operation need only be done mentally; the *abstract polyhedral surface S* consists merely of the triangles T_i, the pairs of associated sides (the *edges* of S), and the vertices.

163.
We shall consider polygonal paths $PP_1 ... P_{n-1}Q$, directed from P to Q, containing a finite number of adjacent edges of S. We call S *connected* if any two of its vertices P and Q can be joined by such a path.

A connected surface is *closed* if it contains only a finite number of triangles, *open* if it contains an infinite number.

Given a path α from P to Q, we shall denote by α^{-1} the same path described from Q to P. If β is a path from Q to R, we shall denote by $\alpha\beta$ the path PQR obtained by describing α from P to Q followed by β from Q to R.

We *deform* a path α by making a finite number of operations of the following two kinds:

(a) An edge of α which corresponds to a side AB of a triangle ABC is replaced by $AC + CB$, or $AC + CB$ is replaced by AB.

(b) A pair of successive edges corresponding to $AB + BA$ is inserted or deleted.

If α can be deformed into β in this way, β is called *homotopic* to α. Clearly this is an equivalence relation, so that the paths joining P to Q are divided into *classes* $\{PQ\}$ of homotopic paths.

For most surfaces, there are several classes $\{PQ\}$, generally an infinite number. If there is only one class $\{PQ\}$, i.e. if all paths joining P to Q are homotopic (for any choice of P and Q), the surface S is called *simply connected*.

164. Now consider a fixed vertex O of S, and two paths α and β joining O to P. The path $\alpha\beta^{-1}$ then begins and ends at O, and α and β are homotopic if and only if $\alpha\beta^{-1}$ is homotopic to the path consisting of the single point O.

It follows easily from this that we can choose a set of closed paths containing O

$$\omega^{(1)}, \omega^{(2)}, \ldots \qquad\qquad \ldots\ldots(164\cdot1)$$

such that, given any paths α and β from O to P, exactly one of the paths

$$\alpha, \omega^{(1)}\alpha, \omega^{(2)}\alpha, \ldots \qquad\qquad \ldots\ldots(164\cdot2)$$

is homotopic to β.

165. The universal covering surface.

Consider two surfaces S and $\bar{S}$, formed with triangles T_j and $\bar{T}_j^{(\nu)}$ respectively. Suppose that each of the triangles $\bar{T}_j^{(1)}, \bar{T}_j^{(2)}, \ldots$ is related to the same triangle T_j, which is called the *projection* of $\bar{T}_j^{(\nu)}$. Suppose further that, if $\bar{T}_j^{(\nu)}$ and $\bar{T}_k^{(\mu)}$ are adjacent on $\bar{S}$, their projections T_j and T_k are adjacent on S. Then $\bar{S}$ is called a *covering surface* of S.

A covering surface $\bar{S}$ of S is called *unbranched* if every cycle of triangles adjoining a vertex $\bar{P}$ of $\bar{S}$ projects on to a similar cycle on S, both cycles having the same number of triangles.

166. We shall now show that any surface S has a *simply connected, unbranched* covering surface $\bar{S}$, called the *universal covering surface* of S.

To construct $\bar{S}$, we first choose a vertex O of S and a set ($164\cdot1$) of closed paths on S containing O. Next, given any triangle T_n of S, with vertices P, Q and R, we associate with it triangles

$$\bar{T}_n, \bar{T}_n^{(1)}, \bar{T}_n^{(2)}, \ldots \qquad\qquad \ldots\ldots(166\cdot1)$$

where $\bar{T}_n^{(\nu)}$ corresponds to the path $\omega^{(\nu)}\alpha$ (or any path joining O to P homotopic to $\omega^{(\nu)}\alpha$). This involves selecting a definite vertex P from each triangle T_n; we restore symmetry by associating with $\bar{T}_n^{(\nu)}$ not only a class $\{\beta\}$ of paths from O to P, but also classes $\{\beta'\}$ and $\{\beta''\}$ of paths from O to Q and O to R. The classes are related in pairs by the rule:

A path β' is homotopic to $\beta + PQ$, a path β'' homotopic to $\beta + PR$.

Finally, we arrange that the vertices $P^{(\nu)}$, $Q^{(\nu)}$, $R^{(\nu)}$ of $\bar{T}_n^{(\nu)}$ project on P, Q, R, and similarly for the sides.

To complete the definition of $\bar{S}$, we must decide how to associate pairs of sides of triangles on $\bar{S}$. Let T_n and T_m be triangles of S with a common edge PQ, and let α be any path from O to P. This path gives rise to well defined triangles $\bar{T}_n{}^{(\nu)}$, $\bar{T}_m{}^{(\mu)}$, and we associate those two sides of $\bar{T}_n{}^{(\nu)}$ and $\bar{T}_m{}^{(\mu)}$ which project on PQ. This makes the collection of triangles $\bar{T}_n{}^{(\nu)}$ into an abstract surface $\bar{S}$ which is clearly an unbranched covering surface of S.

To prove that $\bar{S}$ is simply connected, take any two vertices $\bar{P}$ and $\bar{Q}$ of $\bar{S}$ projecting on P and Q, and let $\bar{\beta}$ and $\bar{\gamma}$ be any two paths joining $\bar{P}$ to $\bar{Q}$ projecting on to β and γ joining P to Q. To $\bar{P}$ corresponds a path α from O to P, and both the paths $\alpha\beta$ and $\alpha\gamma$ joining O to Q must correspond to $\bar{Q}$. (For if $\bar{\beta} = \bar{P}\bar{P}_1 \ldots \bar{P}_{n-1}\bar{Q}$ and $\beta = PP_1 \ldots P_{n-1}Q$, the paths $\alpha + PP_1 \ldots P_i$ and $\alpha + PP_1 \ldots P_{i+1}$ belong to related classes as defined above. It follows easily by induction that $\alpha + P \ldots P_i$ corresponds to $\bar{P}_i$, and in particular that $\alpha\beta$ corresponds to $\bar{Q}$.) Now since $\alpha\beta$ and $\alpha\gamma$ both correspond to $\bar{Q}$, they must be homotopic, and hence so are $\alpha^{-1}\alpha\beta$ and $\alpha^{-1}\alpha\gamma$, and therefore β and γ are homotopic.

But the operations (a) and (b) which deform β into γ will deform $\bar{\beta}$ into $\bar{\gamma}$ if transferred to $\bar{S}$. Thus $\bar{\beta}$ and $\bar{\gamma}$ are homotopic, and $\bar{S}$ is simply connected.

167. Domains and their boundaries.

Given any finite collection $T_{n_1}, T_{n_2}, \ldots, T_{n_k}$ of triangles on S, their sum *taken modulo* 2 will be called a *domain* Σ of S and denoted by

$$\Sigma = T_{n_1} + T_{n_2} + \ldots + T_{n_k}. \qquad \ldots\ldots(167\cdot1)$$

Any finite number of domains $\Sigma_1, \Sigma_2, \ldots, \Sigma_m$ can also be added modulo 2, so that the symbol $\Sigma_1 + \Sigma_2 + \ldots + \Sigma_m$ has a meaning.

The *boundary* $\mathfrak{B}$ (T) of a triangle is the sum of its sides; the boundary of the domain $(167\cdot1)$ is then defined by the equation

$$\mathfrak{B}(\Sigma) = \mathfrak{B}(T_{n_1}) + \mathfrak{B}(T_{n_2}) + \ldots + \mathfrak{B}(T_{n_k}). \quad \ldots\ldots(167\cdot2)$$

This is a very convenient formalism for proving the theorems which follow:

If α and β are homotopic paths, and if $[\alpha]$ and $[\beta]$ are the sums modulo 2 of their sides, then $[\alpha] + [\beta]$ is the boundary $\mathfrak{B}(\Sigma)$ of some domain Σ.

For if a triangle T is used for operation (a) of § 163, $[\alpha]$ is transformed into $[\alpha] + \mathfrak{B}(T)$, whilst operation (b) leaves $[\alpha]$ unchanged. Hence if β is homotopic to α we can write

$$[\beta] = [\alpha] + \mathfrak{B}(T_{n_1}) + \ldots + \mathfrak{B}(T_{n_k}) = [\alpha] + \mathfrak{B}(\Sigma),$$

and therefore

$$[\alpha] + [\beta] = \mathfrak{B}(\Sigma). \qquad \ldots\ldots(167\cdot3)$$

Every non-empty domain Σ *on an open connected surface* S *has a non-empty boundary* $\mathfrak{B}$ (Σ).

Since S is open there exist triangles T of S which do not belong to Σ, and since S is connected at least one such triangle is adjacent to a triangle of Σ.

Two different domains Σ *and* Σ' *on an open connected surface* S *cannot have the same boundary.*

For if $\mathfrak{B}$ $(\Sigma) = \mathfrak{B}$ (Σ'), then $\mathfrak{B}$ $(\Sigma + \Sigma') = 0$, and this contradicts the previous theorem since $\Sigma + \Sigma'$ is not empty. It should be observed that on a *closed* surface there are always pairs of (complementary) domains with the same boundary.

On an open simply connected surface, every simple closed polygon is the boundary of exactly one domain.

Two distinct vertices P and Q of the polygon π divide it into paths α and β from P to Q. Since S is simply connected, α and β are homotopic. Also (since the polygon is simple) $\alpha = [\alpha]$ and $\beta = [\beta]$, and therefore by (167·3)

$$\pi = [\alpha] + [\beta] = \mathfrak{B} \; (\Sigma).$$

168. The Theorem of van der Waerden (30).

Given a simply connected open surface S, *its triangles can be arranged in a sequence*

$$T_1, T_2, ..., T_n, ... \qquad \qquad ......(168\cdot1)$$

in such a way that each domain

$$\Sigma_n = T_1 + T_2 + ... + T_n \quad (n = 1, 2, ...) \quad ......(168\cdot2)$$

has a simple closed polygon as boundary.

Suppose that $T_1, ..., T_n$ have already been chosen. We must then find T_{n+1} adjacent to Σ_n such that $\mathfrak{B}$ $(\Sigma_n + T_{n+1})$ is a simple closed polygon. This will be so if T_{n+1} has one or two sides in common with the polygon $\mathfrak{B}$ (Σ_n) *but not one side and the opposite vertex.* We must also ensure that every triangle of S occurs in the sequence (168·1).

Choose any triangle Δ of S adjacent to Σ_n. The sides p, q and r of Δ cannot all lie on $\mathfrak{B}$ (Σ_n), because if they did $\mathfrak{B}$ (Σ_n) would not be a simple closed polygon. Therefore one of three things must happen:

I. Two sides of Δ lie on $\mathfrak{B}$ (Σ_n).

II. One side of Δ lies on $\mathfrak{B}$ (Σ_n), the opposite vertex does not.

III. One side p of Δ and the opposite vertex P both lie on $\mathfrak{B}$ (Σ_n).

If Δ satisfies I or II, we take $T_{n+1} = \Delta$. In case III, the polygon $\mathfrak{B}(\Sigma_n)$ splits into three parts p, α and β, and $\alpha + q$, $\beta + r$ are simple closed polygons, bounding domains Π and Π^* (see Fig. 45).

Thus

$$\mathfrak{B}(\Pi) = \alpha + q, \quad \mathfrak{B}(\Pi^*) = \beta + r$$

and

$$\mathfrak{B}(\Pi + \Pi^*) = (p + \alpha + \beta) + (p + q + r)$$
$$= \mathfrak{B}(\Sigma_n) + \mathfrak{B}(\Delta).$$

It follows that

$$\Pi + \Pi^* = \Sigma_n + \Delta. \qquad \ldots\ldots(168\cdot3)$$

The triangles $T_1, \ldots, T_n$ belong to Σ_n, and so each of them belongs either to Π or to Π^*. But if T_1 belongs to Π^*, say, then so do $T_2, \ldots, T_n$ and Δ, because no side of the polygon $\mathfrak{B}(\Pi^*) = \beta + r$ belongs to more than one of these $(n + 1)$ triangles. Thus Π consists of triangles outside $\Sigma_n + \Delta$; we now try to choose T_{n+1} from one of these triangles.

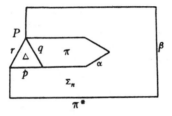

Fig. 45

Let Δ' be a triangle in Π adjacent to α. We repeat the argument above, with Δ replaced by Δ', and find that: either we may choose $T_{n+1} = \Delta'$, or there is a polygon $q' + \alpha'$ bounding a domain Π'. But all the vertices of $q' + \alpha'$ lie in α, and so α' is obtained from α by removing at least one side. We now repeat the argument, starting with a triangle Δ'' in Π' adjacent to α'. After a finite number of steps, we must arrive at a suitable triangle T_{n+1} which, being adjacent to α, belongs to Π.

Now replace, in the original figure, Σ_n by $\Sigma_{n+1} = \Sigma_n + T_{n+1}$ and Π by $\Pi - T_{n+1}$, and repeat the argument. After a finite number of steps, we obtain

$$\Sigma_n + \Pi = \Sigma_n + T_{n+1} + \ldots + T_{n+m} = \Sigma_{n+m}$$

and are now ready to adjoin the triangle $\Delta = T_{n+m+1}$.

To show that S can be exhausted by this process, we start with any triangle T_1' of S, adjoin all adjacent triangles, then adjoin all triangles adjacent to those obtained already, and so on. In this way we obtain a sequence $\{T_n'\}$ of triangles exhausting S. If we now arrange, at each stage of the process described above, that Δ is chosen to be the first suitable triangle T_n', it is clear that every triangle T_n' will appear somewhere in the sequence (168·1).

169. Riemann surfaces.

So far, our arguments have not involved any individual points on S except the vertices; the edges of S may be regarded as ordered pairs of vertices, the triangles as triples of vertices.

In order to obtain a *Riemann surface*, we must convert this framework into a *two-dimensional topological space which carries the local angular metric of the ordinary plane* (31) (32).

This is done most simply by assigning to each triangle T_j a Jordan curve α_j lying in the plane of a complex variable t_j. On α_j we mark three points P_j, Q_j and R_j, and we regard the interior of α_j as a *conformal representation* of T_j. Since any two Jordan domains can be transformed conformally into each other, the transformation being continuous if extended to their frontiers (§ 137), we see that both the *shape* and *size* of α_j, and the positions of P_j, Q_j and R_j, are irrelevant. Nevertheless, we can now identify individual points of T_j: we transform the interior of α_j conformally into a circle, so that any point of T_j is mapped into a point A_j of the circle and is characterised by the cross-ratio

$$(P_j, Q_j, R_j, A_j) \quad \text{(see § 5)}.$$

The angular metric which we want to set up is then well defined at all *inner* points of T_j, i.e. those for which (P_j, Q_j, R_j, A_j) is not real.

We must still define the angular metric on the sides and vertices of all our triangles. To do this, we need further assumptions which are best dealt with by considering a special case.

Suppose that the triangle T_1 with sides p, l, q is to be welded along q to T_2 with sides q, m, r (Fig. 46 a, b). We assume that there is at least one conformal representation of T_1 such that the angular metric of the t_1-plane is valid on the side q of T_1, and that a strip σ between q and γ (Fig. 46a) is the conformal image of a part of T_2 (Fig. 46b). Then the welding of T_1 and T_2 along q can be performed by the method of § 154 (Fig. 46c). This process is uniquely defined if we know the images in the t_1- and t_2-planes of a particular point M of q.

Suppose further that three triangles T_1, T_2 and T_3 form a cycle around a vertex A, and that accordingly T_3 has sides p, r, n (Fig. 46d). We apply the preceding method, with the difference that now the part $(r + A + p)$ of the frontier of T_3 must be welded to the corresponding part of the frontier of $T_1 + T_2$, and that A takes the place of M. The common vertex A lies inside the resulting circle (Fig. 46e), so that the angular metric is certainly defined there.

The case when the cycle around A contains more than three triangles is treated by the obvious extension of the method above.

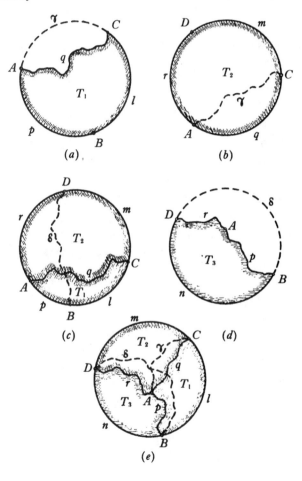

Fig. 46

170. The Uniformisation Theorem.

Let S be a Riemann surface. If S is closed and simply connected, the developments of Ch. VII apply.

In all other cases, the universal covering surface $\bar{S}$ of S is an open, simply connected surface for which the theorem of van der Waerden holds (see § 168). Applying the methods of § 169, we can represent the successive domains

$$\Sigma_n = T_1 + T_2 + \ldots + T_n \quad (n = 2, 3, \ldots)$$

conformally on a circle $|t_n| \leqslant r_n$, and find on this circle the image of Σ_{n-1}. Having done this for all n, we may use the results of §§ 125–130 to establish the

Uniformisation Theorem.

The universal covering surface $\bar{S}$ of any open or closed Riemann surface S can always be represented conformally on

(1) *a closed sphere,*

(2) *an Euclidean plane,*

(3) *the interior of a circle $|t| < R$,*

these possibilities being mutually exclusive.

In this way we obtain a complete *conformal pattern* in the t-plane of the triangulation of $\bar{S}$. Its edges are Jordan curves and do not necessarily have tangents at their end-points, so that in general the angles of the triangles are not defined. However, the given triangulation can always be replaced by an equivalent one for which the edges are analytic arcs or, even more specially, are geodesics in the particular metric involved.

Remark. Care must be taken always to weld adjacent triangles of $\bar{S}$ together in the same way as their projections on S. The resulting triangulation of $\bar{S}$ in the t-plane then admits a group of displacements transforming equivalent triangles into one another.

171. The uniformisation theorem shows that we can always introduce a local metric on $\bar{S}$ which is *spherical* in the first case, *euclidean* in the second and *hyperbolic* (*Lobatschewskyan*) in the third. We can even define the *distance* between any two points of $\bar{S}$ as the distance between their images in the t-plane.

Since two triangles of $\bar{S}$ which have the same projection on S have the same metric at equivalent points, we can transfer the local metric from $\bar{S}$ to S. But we cannot do this for the distance between two points, since in general any point of S has an infinity of corresponding points on $\bar{S}$.

172. Conformal representation of a torus.

It is often important to determine which kind of metric is introduced on a given Riemann surface by the method described above. This can sometimes be done with very little effort.

Consider for instance a torus S in three-dimensional space. By combining the method of § 156 with that of the present chapter, we find that the universal covering surface $\bar{S}$ is represented either on the infinite t-plane or on a finite circle $|t| < R$. The metric induced on S is either *euclidean* or *hyperbolic*.

Fig. 47

We now consider two closed paths α' and β' on S which are homotopic neither to zero nor to each other, and which intersect at a point O of S. The images of α' and β' in the t-plane are arcs joining a point $\bar{O}_1$ to two different points $\bar{O}_2$ and $\bar{O}_3$. We replace α' and β' by curves α and β which are mapped on the segments (euclidean or non-euclidean as the case may be) $\bar{O}_1\bar{O}_2$ and $\bar{O}_1\bar{O}_3$. Since α and β are the shortest closed curves on S passing through O and homotopic to α' and β', they cannot intersect on S except at O.

If we cut S along α and β, we obtain a surface S' which is mapped on a *quadrangle* in the t-plane. This mapping is conformal at O, so that the quadrangle has angle-sum 2π. But the angle-sum of a non-euclidean triangle is less than π, hence the angle-sum of a quadrangle is less than 2π. Therefore the metric must be *euclidean*.

Moreover, opposite sides of the quadrangle have equal length, so that the quadrangle is in fact a *parallelogram*. Any segment parallel to one side and joining two opposite sides is the image of a *closed geodesic* on S. Hence S is covered by two systems of geodesics which form a net on S.

It is well known that an infinity of such nets can be constructed on S. Among them is a "reduced" net, characterised by the fact that the triangle $\bar{O}_1\bar{O}_2\bar{O}_3$ is acute-angled. The shape of the reduced parallelogram is a *conformal invariant* of S: two such surfaces can be transformed conformally into each other if and only if their reduced parallelograms are similar. Since any given point of the t-plane can be moved to another given point by a translation, it follows that any torus S can be transformed conformally into itself in such a way that a given point P is transformed into a second given point Q.

BIBLIOGRAPHICAL NOTES

I. BOOKS

Enzyklopädie der mathematischen Wissenschaften:

L. Lichtenstein. Neuere Entwickelung der Potentialtheorie. Konforme Abbildung. II, c. 3. (Vol. II, 3, 1.)

L. Bieberbach. Neuere Untersuchungen über Funktionen von komplexen Variablen. II, c, 4. (Vol. II, 3, 1.)

H. A. Schwarz. Gesammelte Abhandlungen. II.

G. Darboux. Leçons sur la théorie générale des surfaces. I, Livre II (ch. IV).

E. Picard. Traité d'Analyse. II.

G. Fubini. Introduzione alla Teoria dei Gruppi discontinui e delle Funzioni Automorfe. (Pisa, 1908.)

G. Julia. Leçons sur les fonctions uniformes à point singulier essentiel isolé. (1923.)

—— Principles géométriques d'Analyse. I (1930).

J. L. Coolidge. Treatise on the Circle and the Sphere. (Oxford, 1915.)

Hurwitz-Courant. Funktionentheorie. (3rd ed. 1929.)

H. Weyl. Die Idee der Riemannschen Fläche. (2nd ed. 1922.)

P. Montel. Leçons sur les famillies normales. (Paris, 1927.)

C. Carathéodory. Funktionentheorie. (Basle, 1950.)

II. NOTES AND PAPERS

(1) p. 1. *J. L. Lagrange.* Sur la construction des cartes géographiques. (1779.) Œuvres, IV, pp. 637–92.

(2) p. 2. *C. F. Gauss.* Allgemeine Auflösung der Aufgabe die Theile einer gegebenen Fläche so abzubilden, dass die Abbildung dem abgebildeten in den kleinsten Theilen ähnlich wird. (1822.) Werke, IV, pp. 189–216.

(3) p. 2. *B. Riemann.* Grundlagen für eine allgemeine Theorie der Functionen einer complexen veränderlichen Grösse. (Diss. Göttingen, 1851.) Werke (2nd ed.), pp. 3–41 (§ 21, p. 39).

(4) p. 2. See e.g. *C. Carathéodory.* Variationsrechnung und partielle Differentialgleichungen erster Ordnung (Leipzig, 1935), pp. 305, 309, 333.

(5) p. 2. *D. Hilbert.* Das Dirichletsche Prinzip. (Gött. Festschrift, 1901.) Gesammelte Abhandlungen (Berlin, 1935), III, pp. 15–37.

(6) p. 2. *H. A. Schwarz.* Gesammelte Abhandlungen (Berlin, 1890), II, p. 145. *E. Picard*, Traité d'Analyse, T. II, Chap. X.

(7) p. 4. *A. F. Möbius.* Werke, II, pp. 205–43.

(8) p. 18. *F. Klein.* Vorlesungen über nichteuklidische Geometrie (Berlin, 1928).

(9) p. 18. *H. Poincaré.* Œuvres, II.

(10) p. 22. *P. Finsler.* Über Kurven und Flächen in allgemeinen Räumen. (Diss. Göttingen, 1918.) A reprint (Basle, 1950) is in course of publication.

(11) p. 34. The monodromy theorem was originally given by Weierstrass. The proof he gave in his lectures is to be found in *O. Stolz and J. Gmeiner,* Einleitung in die Funktionentheorie (Leipzig, 1910).

(12) p. 35. Über die Uniformisierung beliebiger analytischen Kurven. Gött.
Nachr. (1907. Paper presented on 12 April.) See especially p. 13.

(13) p. 39. *H. A. Schwarz*. Zur Theorie der Abbildung. (1869.) Gesammelte
Abhandlungen, II, p. 108 (especially § 1).
   The proof given in the text is due to *Erhard Schmidt*. It was first pub-
lished by *C. Carathéodory*. Sur quelques généralisations du théorème de
M. Picard. C.R. 26 Dec. 1905. A similar proof had already been given by
H. Poincaré. Sur les groupes des équations linéaires. Acta Math. Vol. 4
(1884) especially p. 231.

(14) p. 41. *G. Pick*. Über eine Eigenschaft der Konformen Abbildungen
kreisförmiger Bereiche. Math. Annalen, 77 (1916), p. 1.

(15) p. 48. *P. Koebe* made the conjecture that $|f'(0)| \leqq 4$. The first complete
proof was given by *G. Faber*, Neuer Beweis eines Koebe-Bieberbachschen
Satzes über Konforme Abbildung. Münch. Sitzungsber., Math. Phys.
Klasse (1916), p. 39. (Compare *Bieberbach*, Enzyklopädie, l.c. p. 511.)

(16) p. 53. *G. Julia*. Extension nouvelle d'un lemme de Schwarz. Acta Math.
42 (1918), p. 349.
   The general theorem given in the text is due to *J. Wolff* and was published in Comptes Rendus, 13 Sept. 1926. The theorem is here presented
in the form in which it was proved by the author: Über die Winkel-
derivierte von beschränkten analytischen Funktionen. Berl. Sitzungsber.,
Phys. Mathem. Klasse (1929), p. 39, where, in ignorance of Wolff's priority,
his result was proved afresh.

For the definition of continuous convergence for meromorphic functions, cf.

(17) p. 61. *A. Ostrowski*. Über Folgen analytischer Funktionen und einige
Verschärfungen des Picardschen Satzes. Math. Zeitschrift, 24 (1926),
p. 215.

(18) pp. 61, 62. *C. Carathéodory*. Stetige Konvergenz und normale Familien
von Funktionen. Math. Annalen, 101 (1929), p. 315 and l.c. Funktionentheorie, Book IV.

(19) p. 61. Cf. *P. Montel*. l.c.

(20) p. 62. Cf. *M. Fréchet*. Les espaces abstraits. (Paris, 1928.)

(21) p. 66. Three distinct methods of attack have been used to prove the
fundamental theorem of conformal representation: Dirichlet's Principle,
the methods developed in Potential Theory for problems concerning values
on the boundary, and methods taken exclusively from the Theory of
Functions. Proofs on this latter basis were given last of all, after the others
had been disposed of. The history of the problem up to 1918 is exhaustively
dealt with by *Lichtenstein* in his article in the Enzyklopädie.
   The proof presented in this book goes back to *L. Fejér* and *F. Riesz*.
It was published by *T. Radó*. Über die Fundamentalabbildung schlichter
Gebiete. Acta Szeged, I (1923). Radó draws attention to the important
fact that simple connectivity is not made use of in the course of the proof.
For another type of proof, see *C. Carathéodory*, A Proof of the first prin-
cipal Theorem of Conformal Representation, Courant Anniversary Volume
(New York, 1948), and Funktionentheorie, l.c. Book VI, Chap. II.
   Cf. also the interesting paper by *G. Faber*, Über den Hauptsatz der

Theorie der konformen Abbildungen. Münch. Sitzungsber., Math.-Phys. Klasse (1922), p. 91.

See also P. *Simonart*. Sur les transformations ponctuelles et leurs applications géométriques. (2ᵉ partie) la représentation conforme. Ann. de la Soc. Scient. de Bruxelles, 50 (1930), Mémoires, p. 81.

(22) p. 76. The theory also applies to sequences of Riemann surfaces, cf. C. *Carathéodory*. Untersuchungen über die konformen Abbildungen von festen und veränderlichen Gebeiten. Math. Ann. 72 (1912), pp. 107–144.

(23) pp. 85 and 94. The method here used is a simplification of a proof given by E. *Lindelöf*, Sur la représentation conforme d'une aire simplement connexe sur l'aire d'un cercle. 4ᵉ Congrès des mathém. Scandinaves à Stockholm (1916), pp. 59–90. (See especially pp. 75–84.)

The modifications made were suggested in the course of a conversation with T. Radó.

(24) p. 87. For extensions see: E. *Lummer*. Über die konforme Abbildung bizirkularer Kurven vierter Ordnung. Diss. Leipzig (1920).

(25) p. 90. For extensions see: W. *Seidel*. On the distribution of values of bounded analytic functions. Trans. Amer. Math. Soc. 36 (1934), pp. 201–226. C. *Carathéodory*. Zum Schwarzschen Spiegelungsprinzip. Comm. Helvetici, Vol. 19 (1947), pp. 263–278.

(26) p. 97. For extensions see: W. *Seidel*. Über die Ränderzuordnung bei konformen Abbildungen. Math. Annalen, 104 (1931), p. 182.

(27) p. 97. L. *Ahlfors*. Untersuchungen zur Theorie der konformen Abbildungen und der ganzen Funktionen. Acta Fennicae, A, ı, No. 9 (1930).

(28) p. 99. Cf. G. *Darboux*, l.c. ı, Livre 2, Chap. 3.

(29) p. 102. For the application of this theorem to Plateau's problem see: T. *Radó*, The problem of the least area and the problem of Plateau. Math. Zeitschrift, 32 (1930), p. 763.

(30) p. 106. B. L. *van der Waerden*. Topologie und Uniformisierung der Riemannschen Flächen. Leipzig. Sitzungsber. Math.-Phys. Klasse. Vol. 93 (1941).

(31) p. 108. T. *Radó*. Über den Begriff der Riemannschen Fläche. Acta Szeged, 2 (1925), pp. 101–121.

(32) p. 108. C. *Carathéodory*. Bemerkung über die Theorie der Riemannschen Flächen. Math. Zeitschrift, 52 (1950), pp. 703–8.

The §§ 131–136 contain in principle all that is necessary for the study of the representation of the frontier in conformal transformation of the most general simply-connected domain. The relevant literature is mentioned by *Lichtenstein* in his article in the Enzyklopädie (§ 48).

The results of §§ 120–123 on the kernel of a sequence of domains can be extended to the case of representation of 2n-dimensional domains on one another by systems of n analytic functions of n variables. See:

C. *Carathéodory*. Über die Abbildungen die durch Systeme von analytischen Funktionen mit mehreren Veränderlichen erzeugt werden. Math. Zeitschrift, 37 (1932), p. 758.

There is an extensive literature of the representation of multiply-connected domains and Riemann surfaces. (Cf. *Lichtenstein*, l.c. §§ 41, 44 and 45.)

Princeton U-Store, NJ
Sat 30 May 1998
$6.95 + .42 tax = $7.37